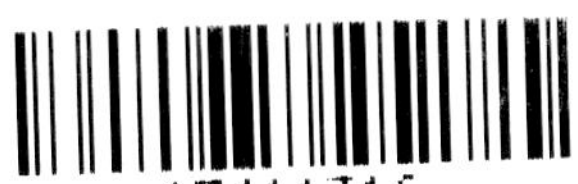
AF411316

Soil Conservation Service Curve Number (SCS-CN) Method Current Applications, Remaining Challenges, and Future Perspectives

Soil Conservation Service Curve Number (SCS-CN) Method Current Applications, Remaining Challenges, and Future Perspectives

Editor

Konstantinos X. Soulis

MDPI • Basel • Beijing • Wuhan • Barcelona • Belgrade • Manchester • Tokyo • Cluj • Tianjin

Editor
Konstantinos X. Soulis
Agricultural University of
Athens
Greece

Editorial Office
MDPI
St. Alban-Anlage 66
4052 Basel, Switzerland

This is a reprint of articles from the Special Issue published online in the open access journal *Water* (ISSN 2073-4441) (available at: https://www.mdpi.com/journal/water/special_issues/Soil_Conservation_Service?authAll=true).

For citation purposes, cite each article independently as indicated on the article page online and as indicated below:

LastName, A.A.; LastName, B.B.; LastName, C.C. Article Title. *Journal Name* **Year**, *Volume Number*, Page Range.

ISBN 978-3-0365-0820-7 (Hbk)
ISBN 978-3-0365-0821-4 (PDF)

Cover image courtesy of Konstantinos Soulis.

Contents

About the Editor . . . . **vii**

Konstantinos X. Soulis
Soil Conservation Service Curve Number (SCS-CN) Method: Current Applications, Remaining
Challenges, and Future Perspectives
Reprinted from: *Water* **2021**, *13*, 192, doi:10.3390/w13020192 . . . . **1**

Muhammad Ajmal, Muhammad Waseem, Dongwook Kim and Tae-Woong Kim
A Pragmatic Slope-Adjusted Curve Number Model to Reduce Uncertainty in Predicting Flood
Runoff from Steep Watersheds
Reprinted from: *Water* **2020**, *12*, 1469, doi:10.3390/w12051469 . . . . **5**

Wenhai Shi and Ni Wang
An Improved SCS-CN Method Incorporating Slope, Soil Moisture, and Storm Duration Factors
for Runoff Prediction
Reprinted from: *Water* **2020**, *12*, 1335, doi:10.3390/w12051335 . . . . **21**

Adam Krajewski, Anna E. Sikorska-Senoner, Agnieszka Hejduk and Leszek Hejduk
Variability of the Initial Abstraction Ratio in an Urban and an Agroforested Catchment
Reprinted from: *Water* **2020**, *12*, 415, doi:10.3390/w12020415 . . . . **39**

Martin Caletka, Monika Šulc Michalková, Petr Karásek and Petr Fučík
Improvement of SCS-CN Initial Abstraction Coefficient in the Czech Republic: A Study of
Five Catchments
Reprinted from: *Water* **2020**, *12*, 1964, doi:10.3390/w12071964 . . . . **55**

Dariusz Młyński, Andrzej Wałęga, Leszek Książek, Jacek Florek and Andrea Petroselli
Possibility of Using Selected Rainfall-Runoff Models for Determining the Design Hydrograph
in Mountainous Catchments: A Case Study in Poland
Reprinted from: *Water* **2020**, *12*, 1450, doi:10.3390/w12051450 . . . . **83**

**Lloyd Ling , Zulkifli Yusop, Wun-She Yap, Wei Lun Tan, Ming Fai Chow and Joan
Lucille Ling**
A Calibrated, Watershed-Specific *SCS-CN* Method: Application to Wangjiaqiao Watershed in
the Three Gorges Area, China
Reprinted from: *Water* **2020**, *12*, 60, doi:10.3390/w12010060 . . . . **101**

Minseok Kang and Chulsang Yoo
Application of the SCS–CN Method to the Hancheon Basin on the Volcanic Jeju Island, Korea
Reprinted from: *Water* **2020**, *12*, 3350, doi:10.3390/w12123350 . . . . **115**

Emmanouil Psomiadis, Konstantinos X. Soulis and Nikolaos Efthimiou
Using SCS-CN and Earth Observation for the Comparative Assessment of the Hydrological
Effect of Gradual and Abrupt Spatiotemporal Land Cover Changes
Reprinted from: *Water* **2020**, *12*, 1386, doi:10.3390/w12051386 . . . . **133**

About the Editor

Konstantinos X. Soulis (Ph.D.) is an Agricultural Engineering scholar with a PhD in Hydrology, GIS and Agricultural Water Management. He works as Laboratory Teaching Personnel, at the Water Resources Division of the Department of Natural Resources Management and Agricultural Engineering, of the Agricultural University of Athens (Greece). His main research and teaching subjects are: Hydrology, Hydrological Modeling, Agricultural Water Management, Geographical Information Systems, Earth Observation, Irrigations, WRM, and Applications Development. He is the author of more than 44 peer-reviewed articles in leading international journals as well as many publications in conference proceedings, books, and professional articles. He has one patent granted and he is a top peer reviewer in Publons. Dr. Soulis serves as an Associate Editor at the *Hydrological Sciences Journal*, as a member of the editorial board of the *Journal of Mountain Science*, and as a Guest Editor in Water Journal and in *ISPRS International Journal of Geo-Information*.

Editorial

Soil Conservation Service Curve Number (SCS-CN) Method: Current Applications, Remaining Challenges, and Future Perspectives

Konstantinos X. Soulis

Department of Natural Resources Management and Agricultural Engineering, Agricultural University of Athens, 75 Iera Odos st., 11855 Athens, Greece; soco@aua.gr or k.soulis@gmail.com

Citation: Soulis, K.X. Soil Conservation Service Curve Number (SCS-CN) Method: Current Applications, Remaining Challenges, and Future Perspectives. *Water* **2021**, *13*, 192. https://doi.org/10.3390/w13020192

Received: 11 January 2021
Accepted: 14 January 2021
Published: 14 January 2021

Publisher's Note: MDPI stays neutral with regard to jurisdictional claims in published maps and institutional affiliations.

Predicting runoff in ungauged or poorly gauged watersheds is one of the key problems in applied hydrology. Thus, simple methods for runoff estimation are particularly important in hydrologic applications, such as flood design or water balance calculation models.

Probably, the most well-documented and at the same time simple conceptual method for predicting runoff depth from rainfall depth is the Soil Conservation Service Curve Number (SCS-CN) method. This method was originally developed by the U.S. Department of Agriculture, Soil Conservation Service (now known as Natural Resources Conservation Service—NRCS) to predict direct runoff volumes for given rainfall events and mainly for the evaluation of storm runoff in small agricultural watersheds. It was initially published in the late fifties and since then has been updated several times with the latest one being in 2004 [1]. Due to its simplicity and its extensive documentation, it soon became a ubiquitous technique that is used worldwide by many engineers and practitioners in numerous hydrological applications. The main reasons behind its impressive success are that it is a very simple but well established method, it features easy to obtain and well-documented environmental inputs, and it accounts for many of the factors affecting runoff generation, incorporating them in a single parameter, the curve number (CN) [2].

Since its creation, the SCS-CN method has been adopted for various regions and for various land uses and climatic conditions, it has been applied to a wide variety of applications beyond its original scope including runoff estimation in large scale river basins and integration in long-term, daily time-step, hydrological models. It has been also the subject of numerous analyses on both practical and theoretical grounds and of several modifications, adaptation, and improvement attempts for over 60 years. Interestingly, after all these attempts, the original formulation of the method is mostly used up to now. Nevertheless, the constantly increasing attention that it receives in the hydrologic literature enhances the current understanding and highlights several critical issues, limitations, and areas for possible improvements, which are still remaining after many years of constant development and research [3].

Accordingly, the aim of this special issue is to present the latest developments in the SCS-CN methodology including but not limited to novel applications, theoretical and conceptual studies broadening the current understanding, studies extending the method's application in other geographical regions or other scientific fields, substantial evaluation studies, and ultimately key advancements towards addressing the key remaining challenges [3] such as:

- Improving the SCS-CN method runoff predictions without sacrificing its current level of simplicity [4–6].
- Moving towards a unique generally accepted procedure for CN determination from rainfall-runoff data considering the prevailing soil and land cover spatial variability in natural watersheds [7–10].

- Improving the initial abstraction estimation and balancing the gains and the implications of altered initial abstraction ratios [11,12].
- Investigating the scale dependency of CN values and the compatibility of CN values obtained at different scales (plot scale, catchment scale, etc.).
- Investigating the integration of SCS-CN method in long-term continuous hydrological models and the implementation of various soil moisture accounting systems [4,13].
- Extending and adopting the existing CNs documentation in a broader range of regions, land uses and climatic conditions including emerging land uses such as roof gardens, solar farms, or porous pavements and considering the effect of gradual and abrupt land cover changes such as urbanization and wildfires [6,10,14–16].
- Utilizing novel modeling, geoinformation systems, and remote sensing techniques to improve the performance and the efficiency of the method [7].

This Special Issue of Water journal comprises eight papers covering many aspects of the above-mentioned challenges [4–7,10–12,14] and spreading in a wide range of different regions and conditions. Among them, two papers are proposing improvements on the SCS-CN method incorporating adjustments for the effects of slope [4,5], initial soil moisture, and storm duration [4]. The long-debated issue of the initial abstraction is investigated in three papers [5,11,12], while three additional papers investigate the application of the method in diverse geographical regions and conditions [6,10,14]. Finally, the utilization of the SCS-CN method along with earth observation for the assessment of the hydrological effect of gradual and abrupt land cover changes is also investigated in one more paper [7].

Specifically, Ajmal et al. [5] examined the initial abstraction ratio (λ) and the watershed slope factor effect in estimating runoff. They proposed a slope-adjusted CN model based on detailed rainfall-runoff data coming from 1779 storm events that took place in 39 watersheds with a varying average slope on the Korean Peninsula. The proposed variants were compared with the original model achieving good agreement between the observed and estimated runoff by one of the proposed variants of the CN model. The proposed variant incorporated slope adjusted CN values and an altered initial abstraction ratio, λ = 0.01. The obtained results also indicated that the effects of rainfall and other runoff-producing factors must be incorporated in CN value estimation to accurately reflect the watershed conditions.

Shi and Wang [4] also attempted to consider the watershed slope as well as the initial moisture conditions and storm duration to improve the performance of the SCS-CN method. The proposed method combines the tabulated CN value with a slope gradient, a soil moisture, and a storm duration factor. It was successfully tested using a detailed dataset coming from three runoff plots in a watershed of the Loess Plateau of China. Additionally, a sensitivity analysis indicated that the most sensitive factor is soil moisture followed by storm duration and slope, while initial abstraction ratio was found to be less sensitive.

Krajewski et al. [11] investigated the variability of the initial abstraction ratio in urban and agroforest land uses using as an example two small Polish watersheds with corresponding land use types (urban and agroforest). To this end, detailed data for the storm events recorded between 2009 and 2017 in the case study watersheds were analyzed to investigate the variability of the initial abstraction ratio across events, seasons, and land use types. The obtained results indicated that λ varied between storm events and seasons and was often lower than the original value of 0.20. The study concluded that optimally λ should be locally verified.

Caletca et al. [12] investigated the applicability of the original SCS-CN method in five representative experimental watersheds in the Czech Republic and attempted to determine appropriate initial abstraction ratio values for the physiographic conditions of Central Europe to improve direct-runoff estimates. The possible influence of individual storm event characteristics was also assessed using principal component and cluster analysis. The obtained results indicated that the original CN method in its traditional arrangement does not perform sufficiently well in the studied watersheds and that watershed depended modifications of λ and CN parameters are required. The need for a systematic yet site-specific revision of the traditional CN method, which may help to improve the SCS-CN method performance was also highlighted.

Młyński et al. [6] systematically analyzed and compared the performance of various CN-based rainfall-runoff models to determine the design hydrograph and the related peak flow in a mountainous watershed. To this end, the observed series of hydrometeorological data for the Grajcarek watershed in Poland for the years 1981 to 2014 were utilized. The obtained results indicated that the EBA4SUB model may be a good alternative in determining the design hydrograph in ungauged mountainous catchments as compared to Snyder and NRCS-UH models. However, it was also emphasized that a decisive factor influencing the performance of rainfall-runoff models is the quality of meteorological data constituting the input signal.

Ling et al. [10] presented a method for the improved watershed-specific calibration of SCS-CN model using non-parametric inferential statistics with rainfall–runoff data pairs. The proposed method generated confidence intervals to determine the optimum parameter values by analyzing the corresponding data. The proposed method was tested in the Wangjiaqiao watershed in China and overperformed the runoff prediction accuracy obtained by the asymptotic curve number fitting method, the linear regression model, and the conventional direct SCS-CN model.

Kang and Yoo [14] tested the applicability of the SCS-CN method in a study area with distinctive characteristics, namely the volcanic Jeju Island, South Korea. Three key issues regarding the application of the SCS-CN method were investigated; i.e., the relationship between the initial abstraction and the maximum potential retention, the determination of the maximum potential retention, and the effect of the antecedent soil moisture condition (AMC) on the initial abstraction and CN, by analyzing several storm events observed in the Hancheon watershed. The obtained results indicated that the optimum accumulated number of days for AMC determination is four or five. The antecedent five-day rainfall amount for the AMC-III condition was found to be higher than 400 mm, and for AMC-I less than 100 mm. These values are overpoweringly higher than the values included in the method documentation signifying the importance of local assessment in that distinctive environment. The optimum λ value was also found to be around 0.3.

Lastly, Psomiadis et al. [7] employed SCS-CN method along with earth observation to make a comparative assessment of the hydrological effect of gradual and abrupt land cover changes using as an example a Mediterranean peri-urban watershed in Attica, Greece. The study area underwent a significant population increase and a rapid increase of urban land uses, from the 1980s to the early 2000s and was also affected by several wildfires. A key observation of this study is that the impact of forest fires is much more prominent in rural watersheds than in urbanized watersheds and that runoff increments due to urbanization seem to be higher than runoff increments due to forest fires affecting the associated hydrological risks. The results of this study indicated that SCS-CN method in conjunction with earth observation is a valuable tool for similar assessments in a simple but efficient way.

In conclusion, this special issue provided the ground for the presentation of the latest developments on SCS-CN method including important steps towards addressing the key challenges and limitations. At the same time, some areas for possible further improvement were also identified and highlighted.

Funding: This research received no external funding.

Institutional Review Board Statement: Not applicable.

Informed Consent Statement: Not applicable.

Data Availability Statement: Data sharing not applicable.

Conflicts of Interest: The authors declare no conflict of interest.

References

1.   NRCS National Engineering Handbook Part 630 Hydrology. Available online: https://www.nrcs.usda.gov/wps/portal/nrcs/detailfull/national/water/manage/hydrology/?cid=stelprdb1043063 (accessed on 5 January 2021).
2.   Ponce, V.M.; Hawkins, R.H. Runoff Curve Number: Has It Reached Maturity? *J. Hydrol. Eng.* **1996**, *1*, 11–19. [CrossRef]
3.   Hawkins, R.H.; Moglen, G.E.; Ward, T.J.; Woodward, D.E. Updating the Curve Number: Task Group Report. In *Proceedings of the Watershed Management 2020: A Clear Vision of Watershed Management—Selected Papers from the Watershed Management Conference 2020*; American Society of Civil Engineers (ASCE): Reston, VA, USA, 2020; pp. 131–140.
4.   Shi, W.; Wang, N. An improved SCS-CN method incorporating slope, soil moisture, and storm duration factors for runoff prediction. *Water* **2020**, *12*, 1335. [CrossRef]
5.   Ajmal, M.; Waseem, M.; Kim, D.; Kim, T.W. A pragmatic slope-adjusted curve number model to reduce uncertainty in predicting flood runoff from steep watersheds. *Water* **2020**, *12*, 1469. [CrossRef]
6.   Młyński, D.; Wałega, A.; Ksiazek, L.; Florek, J.; Petroselli, A. Possibility of using selected rainfall-runoff models for determining the design hydrograph in mountainous catchments: A case study in Poland. *Water* **2020**, *12*, 1450. [CrossRef]
7.   Psomiadis, E.; Soulis, K.X.; Efthimiou, N. Using SCS-CN and earth observation for the comparative assessment of the hydrological effect of gradual and abrupt spatiotemporal land cover changes. *Water* **2020**, *12*, 1386. [CrossRef]
8.   Soulis, K.X.; Valiantzas, J.D. SCS-CN parameter determination using rainfall-runoff data in heterogeneous watersheds—The two-CN system approach. *Hydrol. Earth Syst. Sci.* **2012**, *16*, 1001–1015. [CrossRef]
9.   Soulis, K.X.; Valiantzas, J.D. Identification of the SCS-CN Parameter Spatial Distribution Using Rainfall-Runoff Data in Heterogeneous Watersheds. *Water Resour. Manag.* **2013**, *27*, 1737–1749. [CrossRef]
10.   Ling, L.; Yusop, Z.; Yap, W.S.; Tan, W.L.; Chow, M.F.; Ling, J.L. A calibrated, watershed-specific SCS-CN method: Application to Wangjiaqiao watershed in the three Gorges Area, China. *Water* **2020**, *12*, 60. [CrossRef]
11.   Krajewski, A.; Sikorska-Senoner, A.E.; Hejduk, A.; Hejduk, L. Variability of the initial abstraction ratio in an urban and an agroforested catchment. *Water* **2020**, *12*, 415. [CrossRef]
12.   Caletka, M.; Michalková, M.Š.; Karásek, P.; Fučík, P. Improvement of SCS-CN initial abstraction coefficient in the Czech Republic: A study of five catchments. *Water* **2020**, *12*, 1964. [CrossRef]
13.   Soulis, K.X.; Psomiadis, E.; Londra, P.; Skuras, D. A New Model-Based Approach for the Evaluation of the Net Contribution of the European Union Rural Development Program to the Reduction of Water Abstractions in Agriculture. *Sustainability* **2020**, *12*, 7137. [CrossRef]
14.   Kang, M.; Yoo, C. Application of the SCS–CN Method to the Hancheon Basin on the Volcanic Jeju Island, Korea. *Water* **2020**, *12*, 3350. [CrossRef]
15.   Soulis, K.X. Estimation of SCS Curve Number variation following forest fires. *Hydrol. Sci. J.* **2018**, *63*, 1332–1346. [CrossRef]
16.   Soulis, K.X.; Ntoulas, N.; Nektarios, P.A.; Kargas, G. Runoff reduction from extensive green roofs having different substrate depth and plant cover. *Ecol. Eng.* **2017**, *102*, 80–89. [CrossRef]

Article

A Pragmatic Slope-Adjusted Curve Number Model to Reduce Uncertainty in Predicting Flood Runoff from Steep Watersheds

Muhammad Ajmal [1], Muhammad Waseem [2], Dongwook Kim [3] and Tae-Woong Kim [4],*

[1] Department of Agricultural Engineering, University of Engineering and Technology, Peshawar 25120, Pakistan; engr_ajmal@uetpeshawar.edu.pk

[2] Center of Excellence in Water Resources, University of Engineering and Technology, Lahore 54890, Pakistan; waseem.jatoi@cewre.edu.pk

[3] Department of Civil and Environmental System Engineering, Hanyang University, Seoul 04763, Korea; midas515@hanyang.ac.kr

[4] Department of Civil and Environmental Engineering, Hanyang University, Ansan 15588, Korea

* Correspondence: twkim72@hanyang.ac.kr; Tel.: +82-31-400-5184

Received: 26 April 2020; Accepted: 19 May 2020; Published: 21 May 2020

Abstract: The applicability of the curve number (CN) model to estimate runoff has been a conundrum for years, among other reasons, because it presumes an uncertain fixed initial abstraction coefficient ($\lambda = 0.2$), and because choosing the most suitable watershed CN values is still debated across the globe. Furthermore, the model is widely applied beyond its originally intended purpose. Accordingly, there is a need for more case-specific adjustments of the CN values, especially in steep-slope watersheds with diverse natural environments. This study scrutinized the λ and watershed slope factor effect in estimating runoff. Our proposed slope-adjusted CN ($CN_{II\alpha}$) model used data from 1779 rainstorm–runoff events from 39 watersheds on the Korean Peninsula (1402 for calibration and 377 for validation), with an average slope varying between 7.50% and 53.53%. To capture the agreement between the observed and estimated runoff, the original CN model and its seven variants were evaluated using the root mean square error (RMSE), Nash–Sutcliffe efficiency (NSE), percent bias (PB), and 1:1 plot. The overall lower RMSE, higher NSE, better PB values, and encouraging 1:1 plot demonstrated good agreement between the observed and estimated runoff by one of the proposed variants of the CN model. This plausible goodness-of-fit was possibly due to setting $\lambda = 0.01$ instead of 0.2 or 0.05 and practically sound slope-adjusted CN values to our proposed modifications. For more realistic results, the effects of rainfall and other runoff-producing factors must be incorporated in CN value estimation to accurately reflect the watershed conditions.

Keywords: initial abstraction coefficient; slope-adjusted curve number; rainfall; precise runoff; model accuracy

---

1. Introduction

There is plethora of process-based hydrological models, but they require extensive data, which is a limitation in ungauged watersheds. These process-based models are broadly used to estimate and/or predict hydrologic processes across landscapes and to assess the corresponding impacts of land use/cover changes [1]. Rainfall-runoff modeling is among the most fundamental concepts in hydrology, providing a starting point to estimate flood peaks and design structures. The rainfall-runoff process is a dynamic and complex hydrological phenomenon affected by different physical factors and their interactions [2]. Due to the non-linear relationship between rainfall and runoff, the development of a robust model to predict runoff in ungauged watersheds is difficult and time-consuming [3]. The least complex model that reliably meets the anticipated application is often preferable [4]. The advantages of the Natural Resources Conservation Service (NRCS) curve number (CN) [1] model are its simplicity, predictability, and dependence on only one parameter. The CN model has well-documented data, has been globally tested, and has a rich literature. The CN is a function of soil permeability/infiltration capacity, land use/cover, and other runoff-producing conditions of a watershed; it quantifies direct runoff, requiring only the cumulative rainfall depth and the watershed's CN [5]. The initial abstraction coefficient (λ) and the CN in the CN model are vital to accurately estimate runoff from a watershed [6].

1.1. The CN Model Framework

The CN model is structured to quantify runoff depth (Q) using the cumulative rainstorm depth (P) and maximum potential water retention amount (S), a measure of the ability of a watershed to abstract and retain storm precipitation. Here, P, S, and Q are measured in millimeters.

$$Q = \frac{(P - \lambda S)^2}{P + (1 - \lambda)S} \text{ for } P \geq \lambda S Q = 0 \text{ otherwise} \tag{1}$$

The initial abstraction is the rainstorm depth required before runoff begins. Originally, it was taken as $I_a = \lambda S = 0.2S$; here, S (mm) is related to CN via

$$CN = 100\left(\frac{x}{x + S}\right) \text{ or } S = x\left(\frac{100}{CN} - 1\right) \text{ for } x = 254 \text{ mm (or 10 in)} \tag{2}$$

The dimensionless CN varies from 0 to 100 [5]. Handbook tables for CN selection are based on soil types and land use/land cover. The threshold of $\lambda = 0.2$ is being actively debated across the globe for its inconsistent watershed runoff estimation because $\lambda = 0.05$ has been found to be much more representative [2]. Nevertheless, essentially all handbook CN table values correspond to $\lambda = 0.2$. The corresponding S for $\lambda = 0.05$ is different from that for $\lambda = 0.2$ and, hence, the resulted runoff values are different. The adjustment of CN from $\lambda = 0.2$ to $\lambda = 0.05$ has recently been adopted by the Task Group on Curve Number Hydrology [5], which recommends a new relation as $S_{0.05} = 1.42S_{0.2}$, and leads to

$$CN_{0.05} = \frac{100}{1.42 - 0.0042CN_{0.2}} \tag{3}$$

Several studies have shown considerable differences between handbook-tabulated CN values based on land cover/use and those estimated from watershed observations of rainfall–runoff events [2,5,7–10]. The differences are more prominent with smaller CN values and land types not clearly described in the CN tables [5]. Different studies have evidenced runoff prediction from different biomes using $\lambda < 0.2$ values [2,10–16], suggesting λ in the range of 0.01 to 0.05.

1.2. Effect of Slope on CN and Runoff Estimation

There is no handbook convention but, intuitively, higher-sloped watersheds should have higher CN values. Several CN-based models have documented positive slope-adjustment techniques [10,17–24]. However, some mild negative relationships for limited data are also available [5]. Steep slopes generally give a higher potential for runoff [25], but the impact of slope steepness on runoff generation is a debatable topic. Researchers from different biomes have reported increases in runoff that were attributed to a decrease in infiltration, less detention storage and ponding depth, and high flow velocity [10,19–22,25,26]. Some researchers have captured reduced runoff generation per unit of slope length from steep-slope watersheds with pronounced decreasing storm duration, which might be due to thinning and/or disruption of the crust, differential soil cracking, formation of rills, and more ponding depth [27–33]. However, other studies [34,35] found insignificant effects of slope steepness on runoff. These discrepancies are possibly due to contradiction in experimental settings, as well as land cover and use differences.

To accurately estimate runoff, the CN values found in handbook tables are more effective for rain-fed agricultural watersheds, are less efficient for semi-arid watersheds, and are least successful for forested watersheds [36]. The CN model has a spotty and inconsistent performance history for some forested watersheds (i.e., those in which infiltration potential usually exceeds the rainfall intensities), and for frequent, low-volume, and low-intensity rainfalls. Some researchers found notable problems associated with the tabulated CN values for heavy land cover and humid, forested watersheds, suggesting that the model is inapplicable for runoff estimation in such watersheds [2,9]. For many years, the CN values obtained from handbook tables have been problematic and may need case-specific adjustment when applied in regions with more complex natural environments. The accuracy of the CN value is vital in runoff estimation [37]. The objective of this study was to frame a practically sound slope-adjusted CN equation that could follow the CN theoretical limits (0, 100) and enhance the runoff prediction capability of the CN model from rainstorm events in steep-sloped watersheds.

2. Materials and Methods

2.1. Study Area Description and Climate

South Korea is typical of regions largely influenced by complicated geographical features. Its precipitation patterns have diverse seasonal and regional variability [38]. The elevation (area) of the watersheds included in this study vary from 26 m (42.32 km^2) to 911 m (879.10 km^2) above mean sea level. The average slope of the watershed ranges between 7.50% and 53.53%. The majority of the land cover (about 70.50%) is upland forests, followed by 20.26% agricultural land, urban areas (5.22%), grassland (1.56%), and other land cover distribution (2.45%). The dominant soil types are loam and sandy loam, with some fractions of silt loam. The location of watersheds is shown in Figure 1, and other details can be found in [10].

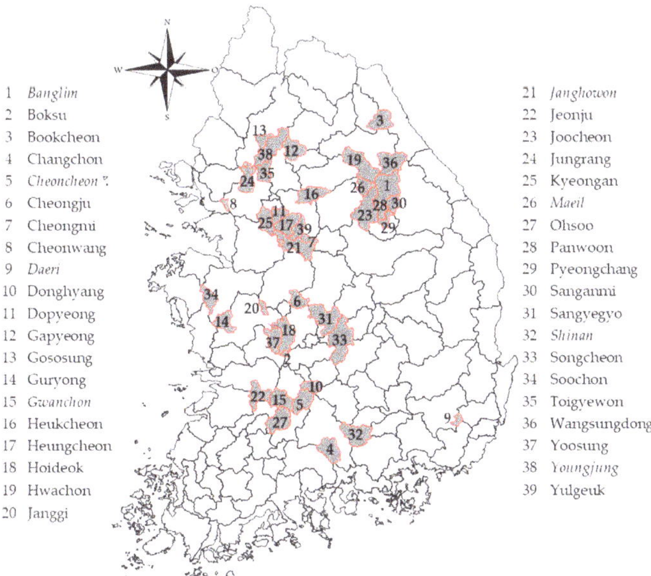

Figure 1. Location of watersheds in the study area. The watersheds in italics were used for validation.

The climatic patterns over the study area are quite variable due to the Asian monsoon. Winter is extremely dry and cold, and summer is warm and moist with frequent heavy rainstorms [38]. The mean annual precipitation (from 1970 to 2000) ranged between 1000 and 1800 mm from the central to the southern regions. Approximately 50% to 60% of this precipitation falls at a high intensity and short duration from July to September [10].

2.2. Data Collection and Interpretation

Continuous rainfall and discharge data (from 2005 to 2012) for this study were collected from the Hydrological Survey Center (HSC) of South Korea. The straight-line hydrograph approach was used to separate direct runoff from the total discharge [10]. For any rain event, the prior five days' cumulative rainfall (P_5) was used to identify the watershed antecedent moisture [10,20,22,39]. The watershed weighted curve number (CN_{II}) corresponding to the normal conditions were derived from the documented tables on the basis of land use/cover and soil types. The CN_I (CN_{III}) for dry (wet) conditions were adjusted as recommended by Mishra et al. [40].

2.3. Slope-Adjusted Curve Number Considerations and Development

Although the CN model is extensively used for predicting runoff from ungauged watersheds, one study found considerable uncertainties when tabulated CN values were applied to estimate runoff from 10 mountainous, forested watersheds in the eastern United States [9]. Similarly, another study [41] observed substantial change in the watershed CN values, ranging from 55 to 70. Moreover, the use of hydrologic soil group D (and its corresponding CN) for forested, mountainous watersheds is incompatible with the National Engineering Handbook [42] guidelines. Although very limited attention has been given to incorporate slope factors in the existing CN models [43], one study reported that adjusting handbook CN values for slope factors significantly enhanced the predicted runoff [26]. To better capture the watershed response in runoff prediction, a slope-adjusted CN is required for steep-slope, mountainous watersheds [10].

Assuming that the handbook CN value is appropriate for a 5% slope [10,17,19,20,22,23], it needs to be adjusted for steep-slope watersheds. To improve the runoff prediction capability of the CN model, the slope-adjusted CN suggested by Sharpley and Williams [17] is generally expressed as

$$CN_{II\alpha} = a(CN_{III} - CN_{II})(1 - be^{-c \times \alpha}) + CN_{II} \tag{4}$$

where $CN_{II\alpha}$ is the slope-adjusted CN for the antecedent runoff condition representing the watershed normal moisture (ARC-II), CN_{II} and CN_{III} are the handbook CN values obtained from watershed characteristics for ARC-II and ARC-III (wet condition), and α is the watershed average soil slope (m/m). The approach of Sharpley and Williams [17] has three empirical parameters—a, b, and c—with optimized values of 1/3, 2, and 13.86, respectively. Their adjusted relationship leads to

$$CN_{II\alpha} = \left(\frac{CN_{III} - CN_{II}}{3}\right)(1 - 2e^{-13.86\alpha}) + CN_{II} \tag{5}$$

Retaining the assumption of Sharpley and Williams [17] for CN_{II} values applicable to a 5% average slope, another study [23] developed the following relationship to adjust CN values for other slopes:

$$S_{II\alpha} = S_{II}\left(1.1 - \frac{\alpha}{\alpha + e^{(3.7 + 0.02117\alpha)}}\right) \tag{6}$$

where S_{II} and $S_{II\alpha}$ are the S values for normal moisture condition and slope-adjusted normal moisture conditions, respectively, and α is the watershed mean slope in percentage. The slope-adjusted CN can be obtained from the above equation using the general S and CN interrelationship as it is found in Equation (2). According to Huang et al. [19], the approach in Sharpley and Williams [17] has not been intensively verified in the field. Hence, they adopted a simplified approach for the $CN_{II\alpha}$ determination on the basis of their experiments for soil slopes ranging between 0.14 and 1.40, and proposed the following relationship:

$$CN_{II\alpha} = CN_{II}\left(\frac{322.79+15.63\alpha}{\alpha+323.52}\right) \tag{7}$$

However, this relationship is unstable because it does not follow the CN theoretical limits.

An investigation by Garg et al. [26] showed that the differences between the tabulated CN values and those calculated from the approach in Huang et al. [19] were very small when compared to that of Sharpley and Williams [17]. This is why the approach in Huang et al. [19] depicted modest improvement in estimating large as well as small runoff events and produced results very close to the original CN model with handbook CN values. Any underestimation of the runoff events using the approach in Huang et al. [19] can be attributed to the empirically selected numerical constants of Equation (7), and needs validation using the measured rainfall-runoff data.

In another study, Ajmal et al. [10] developed a slope-adjusted average CN relationship using data from 39 mountainous watersheds. They calibrated the $CN_{II\alpha}$ using 1402 measured rainfall-runoff events from 31 watersheds and validated this with 377 rainfall–runoff events from the remaining eight watersheds. This is represented as

$$CN_{II\alpha} = CN_{II}\left[\frac{1.9274\alpha+2.13273}{\alpha+2.1791}\right] \tag{8}$$

The above relationship was derived on the basis of data from watersheds with an average slope between 7.50% and 53.53%, where, besides other typical watershed geophysical characteristics, most of the area (approximately 70.50%) was covered with upland forests. However, their approach was also inconsistent with the CN theoretical limits on the basis of the presumption that the CN tables were originally developed with a 5% average slope in their experimental plots [10,17,19]. Knowing CN_{II}, CN_{III}, and α as the mean slope of a watershed, the proposed slope-adjusted CN ($CN_{II\alpha}$) in its general form is presented as

$$CN_{II\alpha} = \left(\frac{CN_{III} - CN_{II}}{2}\right)\left[1 - e^{-b\times(\alpha-0.05)}\right]+CN_{II} \tag{9}$$

2.4. Steps of Slope-Adjusted CN Parameter Optimization

1. Data pertaining to 39 watersheds in which 1779 rainstorms events occurred provided the known values of the rainstorm events, P; the observed runoff, Q_o; and the optimized CNs for each watershed. The least squares nonlinear orthogonal distance regression objective function in Origin Pro 9.6 software produced the optimized CN values from the following equation.

$$\sum_{i=1}^{n}(Q_o - Q_e)^2 = \sum\left\{Q_o - \left[\frac{\left(P - 0.2\times\left(\frac{25400}{CN} - 254\right)\right)^2}{P+0.8\times\left(\frac{25400}{CN} - 254\right)}\right]\right\}^2 = \text{Minimum} \tag{10}$$

2. To optimize parameter b in Equation (9), the CNs obtained for the 39 watersheds from Equation (10) were divided into two sets, those of 31 watersheds (1402 rainstorm–runoff events) for calibration and those of 8 watersheds (377 rainstorm-runoff events) for validation. For calibration, the optimized CNs in step 1 were set as the target values challenging the right side of Equation (9) using the nonlinear regression least squares Levenberg–Marquardt algorithm in SPSS v.25 software. To take into account the individual watersheds' effects on parameter b optimization, the leave-one-out (LOOV) technique was adopted. The average of 31 calibrations repetitions was the value of b = 7.125. This led to recasting the proposed $CN_{II\alpha}$ as

$$CN_{II\alpha} = \left(\frac{CN_{III} - CN_{II}}{2}\right)\left[1 - e^{-7.125\times(\alpha-0.05)}\right]+CN_{II} \tag{11}$$

This can also be represented as

$$CN_{II\alpha} = (0.5 - 0.714e^{-7.125\alpha})(CN_{III} - CN_{II}) + CN_{II} \tag{12}$$

Introducing the CN_{III} conversion from CN_{II} after a suggestion in Mishra et al. [40] gives

$$CN_{III} = \frac{CN_{II}}{0.430 + 0.0057CN_{II}} \tag{13}$$

Imputing Equation (13) into Equation (11) and simplifying it, the proposed relationship can be recast as

$$CN_{II\alpha} = \left[\frac{CN_{II}(50 - 0.5CN_{II})}{CN_{II} + 75.43}\right] \times \left[1 - e^{-7.125(\alpha - 0.05)}\right] + CN_{II} \tag{14}$$

This proposed $CN_{II\alpha}$ relationship has twofold advantages over the previous three suggested relationships. The proposed model has only one parameter to be optimized compared to three in Sharpley and Williams [17] and Williams and Izaurralde [23], and two in Huang et al. [19], if the suggested parameter values are not applicable. Our proposed $CN_{II\alpha}$ works within the theoretical limits (i.e., 0 to 100), unlike that in Huang et al. [19], which loses its effectiveness after $CN_{II} = 94.27$ using the highest average slope of their watersheds. Similarly, the adjustment in Williams and Izaurralde [23] and Ajmal et al. [10] also fails to follow the CN theoretical limits. The different variants of the CN model are shown in Table 1.

Table 1. Models and their descriptions.

Model Identity	λ	CN ($CN_{II\alpha}$)	Model Expression
		Parameters	
M1	0.20	*NEH-4 Tables	Equations (1) and (2)
M2	0.05	NEH-4 Tables	Equations (1)–(3)
M3	0.20	Sharpley and Williams [17]	Equations (1), (2) and (5)
M4	0.20	Huang et al. [19]	Equations (1), (2) and (7)
M5	0.20	Ajmal et al. [10]	Equations (1), (2) and (8)
M6	0.20	Proposed	Equations (1), (2) and (12)
M7	0.05	Proposed	Equations (1)–(3) and (12)
M8	0.01	Proposed	Equations (1), (2) and (12)

*NEH-4: National Engineering Handbook Section-4 [42].

3. Statistical Analysis for Model Performance Evaluation

This study estimated the agreement between a series of observed and estimated runoffs using the root mean square error (RMSE), Nash–Sutcliffe efficiency (NSE), percent bias (PB) [34], and/or graphical assessments augmented with model performance ratings [44]. Mathematically, these indicators are

$$RMSE = \sqrt{\frac{1}{n}\sum_{i=1}^{n}(Q_{oi} - Q_{ei})^2} \tag{15}$$

$$NSE = 1 - \left[\frac{\sum_{I=1}^{n}(Q_{oi} - Q_{ei})^2}{\sum_{I=1}^{n}(Q_{oi} - \overline{Q}_o)^2}\right] \tag{16}$$

$$PB = \left[\frac{\sum_{I=1}^{n}(Q_{oi} - Q_{ei})}{\sum_{I=1}^{n}Q_{oi}}\right] \times 100 \tag{17}$$

where Q_{oi} and Q_{ei} are the observed and estimated runoff values for rainstorm events 1 to n, and $\overline{Q_O}$ is the mean observed runoff in each watershed. The RMSE (0 to ∞) values closer to zero depict more appropriateness of the model to estimate runoff. The NSE ($-\infty$ to 1) illustrates how well a plot of observed vs. estimated runoff fits a 1:1 line (i.e., a perfect fit) [39]. The PB (optimum = 0) describes the average tendency of estimated values to be larger or smaller than their observed ones. Positive (negative) values indicate underestimation (overestimation) bias [44]. It is notable that perfect agreement of the estimated vs. observed data does not essentially indicate a perfect model, because observed data could have uncertainties [39]. However, we are confident about the good quality of the data used in this study. Performance evaluation of different statistical indicators and their suggested ratings [44,45] are given Table 2.

Table 2. Statistical indicators and associated performance ratings [44,45].

Performance Rating	NSE [44]	NSE [45]	PB (%)
Very good	$0.75 < \text{NSE} \leq 1.00$	$0.90 < \text{NSE} \leq 1.00$	$-10 < \text{PB} < +10$
Good	$0.65 < \text{NSE} \leq 0.75$	$0.80 \leq \text{NSE} \leq 0.90$	$\pm 10 \leq \text{PB} < \pm 15$
Satisfactory	$0.50 < \text{NSE} \leq 0.65$	$0.65 \leq \text{NSE} < 0.80$	$\pm 15 \leq \text{PB} < \pm 25$
Unsatisfactory	$\text{NSE} \leq 0.50$	$\text{NSE} \leq 0.65$	$\text{PB} \geq \pm 25$

4. Results and Discussion

The performance evaluation of the existing models (M1–M5) and our proposed approach (M6–M8) was accomplished in two steps. First, the basic statistics of the observed runoff were compared to the models' estimated runoff both for the calibration and validation watersheds. In the second step, commonly used statistical indicators were used to check the model's predictive credibility [20,34,44] in conjunction with a 1:1 plot graphical judgement between the observed and modeled runoff values [46].

4.1. Models' Analysis Based on Descriptive Statistics

The basic descriptive statistics (Table 3) favor the M8 model using the $CN_{II\alpha}$ and lower $\lambda = 0.01$ followed by the M6 and M5 models. However, the M6 model was preferred over the M5 due to its practically sound $CN_{II\alpha}$ to follow the CN theoretical bounds (0–100). In estimating runoff, the M2 model was not plausibly different from the M1 model. Therefore, lowering λ from 0.2 to 0.05, along with its corresponding CN adjustment using Equation (3), produced only modest changes in the estimated runoff values. Nonetheless, using $\lambda = 0.05$ and retaining handbook CN values without adjustment can improve the model's runoff predictive capability, which is not shown in the assessment but is reflected in the comparison of the M6 and M7 models. The majority of the existing CN model variants underestimated the runoff in different watersheds. Nevertheless, it can be inferred that the watershed CN was not the only important parameter; selecting the proper λ also played a crucial role in estimating accurate runoff. Additionally, the prominent response of CNs to the rainstorm depth was vital in runoff depth estimation [1].

Table 3. Summary statistic of rainfall (P), observed runoff (Q_o), and modeled runoff (M1–M8) in the calibration and validation watersheds.

Parameter/Model	Mean	Minimum	First Quartile (Q1)	Median	Third Quartile (Q3)	Maximum
Calibration Watersheds (1402 Rainstorm–Runoff Events)						
P	80.96	12.10	39.92	59.09	98.27	519.68
Q_o	**38.60**	**0.17**	**8.23**	**19.61**	**49.04**	**348.46**
M1	25.57	0.00	1.49	6.13	27.03	415.63
M2	23.56	0.00	1.14	7.26	25.79	**383.27**
M3	28.79	0.00	1.30	7.95	32.94	436.28
M4	26.06	0.00	1.52	6.31	28.33	419.65
M5	30.06	0.00	1.35	8.83	35.39	443.28
M6	30.26	0.00	1.23	9.38	35.34	445.73
M7	28.98	0.00	2.54	10.77	34.57	417.11
M8	**39.67**	**0.53**	**7.93**	**20.13**	**49.30**	**458.55**

Table 3. *Cont.*

	Validation Watersheds (377 Rainstorm–Runoff Events)					
P	75.22	20.52	40.97	57.05	86.95	376.86
Q_0	**35.03**	**0.24**	**8.30**	**19.10**	**43.20**	**364.38**
M1	22.04	0.00	1.48	6.35	20.35	294.27
M2	19.85	0.00	0.85	5.55	19.93	265.59
M3	24.75	0.00	1.52	6.27	25.99	309.31
M4	22.49	0.00	1.39	6.63	21.48	296.26
M5	26.48	0.00	2.03	7.87	30.12	309.72
M6	26.07	0.00	1.71	6.66	29.04	314.48
M7	24.98	0.00	2.10	9.43	26.71	293.91
M8	**34.77**	**0.87**	**7.70**	**17.91**	**40.12**	**325.07**

Note: The highlighted values show the good agreement between the observed and the estimated runoff.

4.2. Model Performance Evaluation in Watersheds Used for Calibration

We evaluated the runoff predictability performance of the existing CN models (M1 to M5) and the proposed variants (M6 to M8) for the calibration watersheds (Figure 2). Because of minimal difference in the $CN_{II\alpha}$ values proposed by Williams and Izaurralde [23] and Sharpley and Williams [17], we compared only the latter with the other approaches. As mentioned earlier, the RMSE can vary from 0 to ∞, and a value close to zero indicates a nearly perfect fit [15,20,34]. On the basis of the RMSE (mean, median) values, the M2 (23.90, 21.91) and M3 (24.30, 21.90) models exhibited similar but improved runoff estimation compared to the M1 (26.49, 24.02) model. The mean value for all of the statistical indicators is shown on each box plot through connected lines. The M2 model's enhanced runoff estimation could be attributed to the lower $\lambda = 0.05$ [2], whereas the M3 model's improved predictability could be ascribed to $CN_{II\alpha}$, which was comparatively higher than the tabulated CN [17]. The M4 model (26.08, 23.78) showed almost no improvement compared to the M1 model. Comparatively better runoff prediction was found for the M5 model (23.53, 21.15), and that of the M6 model (23.23, 20.79) was almost equal in the calibration watersheds. However, the runoff predictive capabilities of the M7 model (21.06, 19.29) and M8 model (18.59, 16.87) were better, as was also evident from their overall RMSE values (Figure 2a). It can be inferred that setting a lower λ and a comparatively higher $CN_{II\alpha}$, as was the case in model M8, possibly reduces the infiltration and surface water retention capacity.

Following the model performance ratings shown in Table 2 and the box plot statistics (Figure 2b), the NSE (mean, median) for the M1 model (0.58, 0.63) and the M4 model (0.59, 0.64) were the smallest among the eight variants of the CN model. It must be kept in mind that the Gusosung watershed statistics were excluded, meaning the mean and median values were calculated for the remaining 30 calibration watersheds. In that particular watershed, only the M8 model showed a reasonable runoff prediction, whereas the rest of the models' performance indicators ratings were unsatisfactory. The M3 model (0.64, 0.68) results showed modest improvement, followed by the M2 (0.66, 0.71) and M5 (0.66, 0.71) models. However, the M6 (0.67, 0.72) and M7 (0.74, 0.77) models exhibited significantly improved results compared to the M1 model. In addition, the M8 model (0.80, 0.82) outperformed all the other models in the majority of the watersheds. The best performance of the M8 model is also evident from Figure 2b, followed by the M7 and M6 models, in that order. The lack of effectiveness of the M1 and M4 models could be attributed to the fixed and higher $\lambda = 0.2$ and inconsistent watershed tabulated CN values [10,15]. Similarly, on the basis of the PB performance ratings (Table 2), the accuracy runoff predictability of the different CN model variants is shown in Figure 2c. Using PB (mean, median), the order for accurately estimating runoff was M8 (−2.43, 0.67) > M7 (19.47, 18.06) > M6 (22.37, 22.51) > M5 (23.22, 21.93) > M3 (25.93, 24.46) > M2 (31.86, 31.26) > M4 (32.93, 32.41) > M1 (34.19, 33.14). In addition, Figure 2c shows that the PB values obtained from the M8 model in estimating runoff in the study area, except for two watersheds, were rated either very good, good, or at least satisfactory.

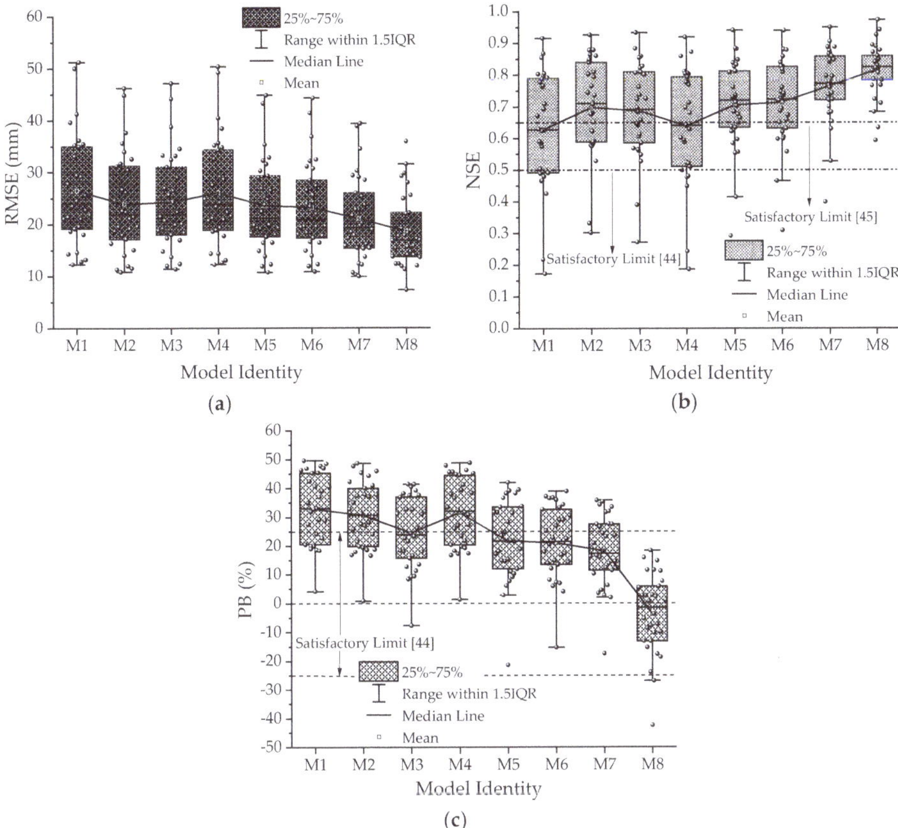

Figure 2. (**a**) Root mean square error (RMSE), (**b**) Nash–Sutcliffe efficiency (NSE), and (**c**) percent bias (PB) for eight variants of the CN model using data of 30 out of 31 calibration watersheds.

4.3. Models' Performance Evaluation in Watersheds Used for Validation

The performance of the CN model variants in the validation watersheds using the RMSE, NSE, and PB is shown in Figure 3. The superior performance of the M8 model is evident, whereas the least efficient was the M1 model with its RMSE, NSE, and PB (mean, median) values of (24.56, 22.73), (0.57, 0.60), and (36.73, 33.18), respectively. The corresponding best runoff prediction by the M8 model was recorded with RMSE (17.25, 16.07), NSE (0.80, 0.78), and PB (−0.35, −3.35). Similarly, the higher PB positive values by the M1 model in the majority of the watersheds indicated underestimation and were in the unsatisfactory range, as found by other researchers [10,20,34,44]. Nevertheless, the M8 model overestimated runoff in the majority of the watersheds, but, was within the acceptable performance range. In addition, among the remaining six variants of the CN model, the M7 model predicted more accurate runoff, followed by the M5, M6, M2, M3, and M4 models, in that order. On the basis of the PB values (Figure 3), the M8 model predicted runoff well in all the watersheds except one.

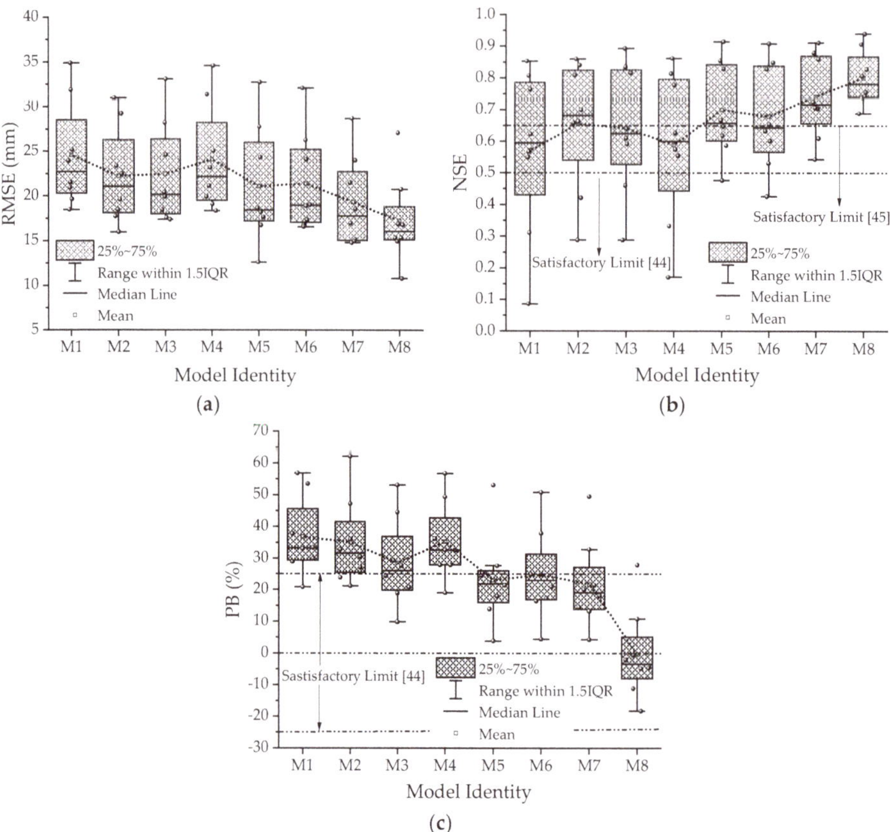

Figure 3. (**a**) RMSE, (**b**) NSE, and (**c**) PB for eight variants of the CN model using data of eight validation watersheds.

4.4. Overall Performance of Models and Comparison Based on 1:1 Plot

Table 4 summarizes the credibility of the eight variants of the CN model in estimating runoff from rainstorm events in different watersheds. It is obvious that the M8 model exhibited more accurate results for a very good performance rating based on NSE (PB) in 30 (19) out of 39 watersheds. The corresponding goodness-of-fit ratting for the M1 model was found only in 14 (1) watershed(s). Applying the model evaluation criteria recommended by Ritter and Muñoz-Carpena [45], the M1 and M4 model predictions were "satisfactorily" to "very good" in only 43.6% of the watersheds, followed by the M3, M5, M2, M6, and M7 models with their corresponding values of 53.9%, 61.5%, 64.1%, 66.7%, and 84.6% of the watersheds, respectively. The more plausible model for efficiently predicting runoff was M8 in 92.3% (36 out of 39) watersheds. It is notable that the majority of the runoff was underestimated by the M1 model, as has also been reported for rangeland and cropland in Montana and Wyoming [47], Mississippi [48], the Loess Plateau of China [19], India [20,22,26,43], South Korea [10,15], and Poland [49]. After M8, the M7 and M6 models predicted runoff more coincident with the observed values. The M4 model's inferior performance could possibly be linked to very little difference in the $CN_{II\alpha}$ and the handbook CN values ($CN_{II\alpha}$–CN), which varied in the range of 0.73 to 1.46. The corresponding CN differences for the M3, M5, and M6 models were in the range of 1.37 to 6.52, 0.73 to 11.28, and 1.15 to 9.48, respectively. It is notable that the M6 and M8 models used the same $CN_{II\alpha}$ values. The M8 model's outperformance in predicting runoff was probably because of its lower $\lambda = 0.01$, as suggested for Korean steep-slope watersheds [10], and its comparatively higher $CN_{II\alpha}$ values.

Table 4. Performance of the CN model and its variants in 39 watersheds in the study area.

	M1	M2	M3	M4	M5	M6	M7	M8
Performance Criteria				**NSE [44]**				
$0.75 < \text{NSE} \leq 1.00$	14	15	14	14	14	14	20	30
$0.65 < \text{NSE} \leq 0.75$	3	10	7	3	10	12	13	6
$0.50 < \text{NSE} \leq 0.65$	10	9	13	13	11	9	4	2
$\text{NSE} \leq 0.50$	12	5	5	9	4	4	2	1
				NSE [45]				
$0.90 < \text{NSE} \leq 1.00$	1	1	1	1	2	2	3	5
$0.80 \leq \text{NSE} \leq 0.90$	6	12	12	8	11	11	11	20
$0.65 \leq \text{NSE} < 0.80$	10	12	8	8	11	13	19	11
$\text{NSE} \leq 0.65$	22	14	18	22	15	13	6	3
				PB (%)				
$-10 < \text{PB} < +10$	1	1	5	1	5	6	6	19
$\pm10 \leq \text{PB} < \pm15$	0	0	3	0	6	5	8	9
$\pm15 \leq \text{PB} < \pm25$	10	11	12	10	13	12	12	7
$\text{PB} \geq \pm25$	28	27	19	28	15	16	13	4

We further compared the different CN model variants on the basis of cumulative observed and estimated runoff from the 39 watersheds using the 1:1 plot and the coefficient of determination, R^2. The moderately high R^2 value supported better runoff prediction capability of the M2 model compared to the M1 model. However, deviation of the observed–estimated runoff best-fit-regression line from the 1:1 plot shows that both the M1 and M2 models underestimated the majority of the runoff events (Figure 4). Although the M2 model R^2 value was comparatively high, the runoff predictability of the M1, M2, and M4 models was almost indistinguishable. Nevertheless, the closeness of data points around the 1:1 plot and the higher R^2 values of the M5 through M8 models favored these models for comparatively better runoff prediction. The best agreement between the observed and estimated runoff was evidenced by applying the M8 model, as shown in Figure 4. It should be noted that the R^2 statistics used for model evaluation could mislead practitioners. These statistics are oversensitive to extremely high values and insensitive to additive and proportional differences between model predictions and measured data [44]. The overall promising results of the M8 model support its suitability for runoff prediction in the steep-slope watersheds. Therefore, the original CN model and the majority of its variants discussed here do not well represent complex watershed characteristics, and thus the abstraction coefficient, the CN values from watershed, and the CN model itself need to be revised for general application. A very recent and comprehensive review by the NRCS Task Group on Curve Number Hydrology [5] also suggested changes to update the handbook and its associated procedures on the basis of lessons learned from global experiences and additional data analyses. To avoid jumps in runoff estimation, the CN model could be made to be more robust by not fixing the initial abstraction coefficient and considering the effect of rainfall as well as the spatial and temporal variability while estimating the watershed CN values.

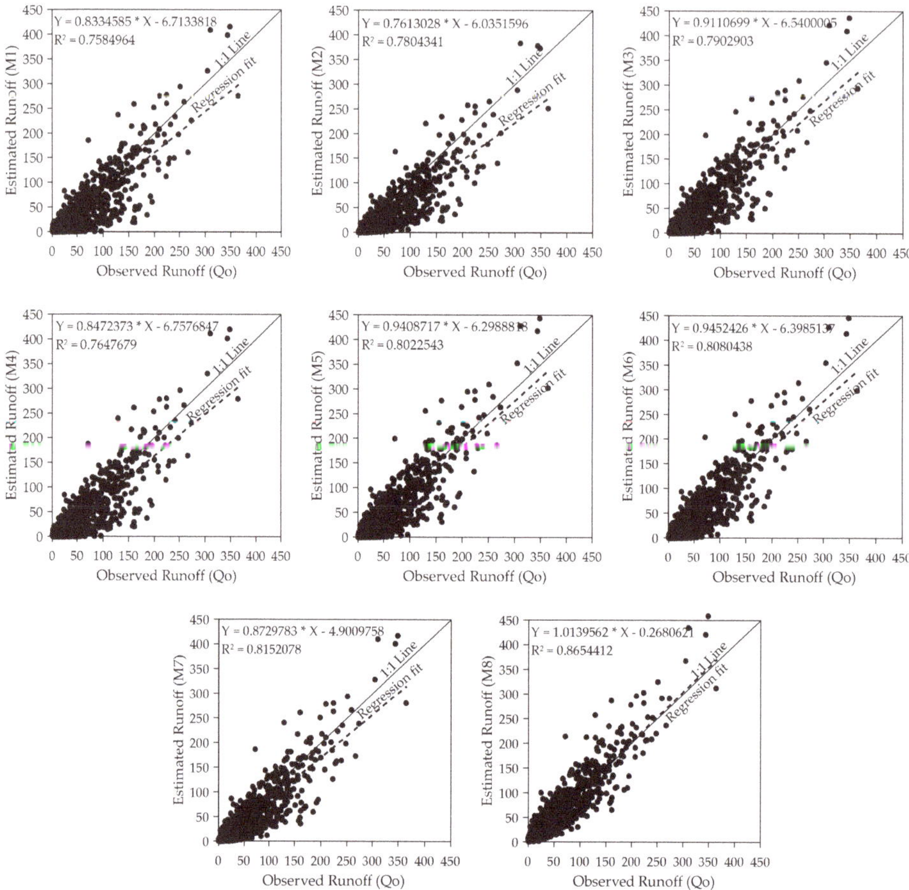

Figure 4. Observed and estimated runoff comparison for eight variants of the CN model using cumulative data of all 39 watersheds.

There is an evidence that the CN tables that were documented a few decades back that were based on soils and land use/cover are often wide of the mark and not supported by real ground data or by critical analyses [10,15,50]. The original CN model response demonstrated in different studies is very sensitive in selecting the watershed-representative CN. Moreover, the runoff response from some watersheds were found to be very erratic, leading to great discrepancies between the modeled data and reality [50]. Like our findings, various studies have reported underestimated runoff in the steep-slope watersheds using the original CN methodology [10,17–23], and slope adjustment for CN was proposed to capture the watershed response in predicting runoff [10,17–19,21–24]. Application of the suggested approach by Sharpley and Williams [17] was criticized for being tested with very limited data in the field [19]. To support the findings of Williams et al. [18], two other slope-adjusted CN approaches were developed by Ajmal et al. [10] and Sharpley and Williams [17], but they were not structurally sound due to incapability to follow the CN theoretical limits. Because of the plausible response in replicating the watershed runoff, the slope-adjusted CN approach proposed in this study was not only structurally sound in terms of following the theoretical bounds of the CN, but also in supporting its application for better runoff prediction. However, the model results could be further improved by introducing the effects of spatial variability in CN for the soil–cover complex along watersheds [51,52].

5. Conclusions and Practical Implications

The CN model is being updated continuously on the basis of new measured rainfall–runoff data and innovation in research. When handbook CN values are used, the inconsistent runoff prediction capability of this model has led researchers to adjust the CN values using the effect of rainfall magnitudes [2,5] and watershed slope [10,17–19,24,26]. However, some researchers agree that the handbook CN values are fit for runoff estimation from watersheds with a maximum 5% average slope. Hence, there is a room for further refinement in determining CN values. This study investigated and proposed a practically sound slope-adjusted CN ($CN_{II\alpha}$) approach to improve the runoff prediction capability of the CN model in steep-slope watersheds in order to reduce possible uncertainties. The proposed $CN_{II\alpha}$ not only followed the theoretical limits (0, 100) [17], but in addition, unlike other existing $CN_{II\alpha}$ approaches [10,19,23], it provided a promising runoff prediction capability in the study area. The use of $\lambda = 0.05$ in place of $\lambda = 0.2$ and their adjusted $CN_{0.05}$ values modestly improved the CN model runoff predictability, but not well enough for runoff estimation from steep-slope watersheds. On the basis of different performance indicators, we found that the proposed $CN_{II\alpha}$ had a positive impact on the CN model runoff prediction. Users of the CN model should know the limitations in its procedures and assumptions because the model produces diverse responses when applied to different land types and watersheds [5]. Assuming a fixed λ value and its associated three fixed values of initial abstraction for dry, normal, and wet conditions are among the major limitations of the original CN model and variants used in this study. The model needs an overhaul for various compelling reasons to circumvent the fixed λ value, as well as unjustified sudden jumps in CN values and its associated estimated runoff. In this era of cutting-edge technology, researchers of different biomes have introduced new parameters in the model to improve its runoff prediction capability. However, inculcating new parameters has increased the model complexity and restricted its application in ungauged watersheds. The CN methodology must be overhauled using experiences from the modern hydrologic engineering without losing the simplicity rule.

Author Contributions: Conceptualization, methodology, and writing—original draft, M.A.; validation, M.W.; visualization and writing—review and editing, D.K.; project administration and resources, T.-W.K. All authors have read and agreed to the published version of the manuscript.

Funding: This research was supported by a grant (2020-MOIS33-006) from the Lower-level and Core Disaster-Safety Technology Development Program funded by the Ministry of Interior and Safety (MOIS, Korea).

Conflicts of Interest: The authors declare no conflict of interest.

References

1. Muche, M.E.; Hutchinson, S.L.; Hutchinson, J.S.; Johnston, J.M. Phenology-adjusted dynamic curve number for improved hydrologic modeling. *J. Environ. Manag.* **2019**, *235*, 403–413. [CrossRef] [PubMed]
2. Hawkins, R.H.; Ward, T.J.; Woodward, D.E.; Van Mullem, J.A. *ASCE EWRI Task Committee Report on State of the Practice in Curve Number Hydrology*; ASCE: Reston, VA, USA, 2009; Volume 106.
3. Fan, F.; Deng, Y.; Hu, X.; Weng, Q. Estimating composite curve number using an improved SCS-CN method with remotely sensed variables in Guangzhou, China. *Remote Sens.* **2013**, *5*, 1425–1438. [CrossRef]
4. Beck, H.E.; van Dijk, A.I.; de Roo, A.; Dutra, E.; Fink, G.; Orth, R.; Schellekens, J. Global evaluation of runoff from ten state-of-the-art hydrological models. *Hydrol. Earth Sys. Sci.* **2017**, *21*, 2881–2903. [CrossRef]
5. Hawkins, R.H.; Theurer, F.D.; Rezaeianzadeh, M. Understanding the basis of the curve number method for watershed models and TMDLs. *J. Hydrol. Eng.* **2019**, *24*, 06019003. [CrossRef]
6. Ling, L.; Yusop, Z.; Chow, M.F. Urban flood depth estimate with a new calibrated curve number runoff prediction model. *IEEE Access* **2020**, *8*, 10915–10923. [CrossRef]
7. Pandit, A.; Heck, H.H. Estimations of soil conservation service curve numbers for concrete and asphalt. *J. Hydrol. Eng.* **2009**, *14*, 335–345. [CrossRef]
8. Stewart, D.; Canfield, E.; Hawkins, R. Curve number determination methods and uncertainty in hydrologic soil groups from semiarid watershed data. *J. Hydrol. Eng.* **2012**, *17*, 1180–1187. [CrossRef]

9. Tedela, N.H.; McCutcheon, S.C.; Rasmussen, T.C.; Hawkins, R.H.; Swank, W.T.; Campbell, J.L.; Adams, M.B.; Jackson, C.R.; Tollner, E.W. Runoff curve numbers for 10 small forested watersheds in the mountains of the eastern United States. *J. Hydrol. Eng.* **2012**, *17*, 1188–1198. [CrossRef]

10. Ajmal, M.; Waseem, M.; Ahn, J.-H.; Kim, T.-W. Runoff estimation using the NRCS slope-adjusted curve number in mountainous watersheds. *J. Irrig. Drain. Eng.* **2016**, *142*, 04016002. [CrossRef]

11. Fu, S.; Zhang, G.; Wang, N.; Luo, L. Initial abstraction ratio in the SCS-CN method in the Loess Plateau of China. *Trans. ASABE* **2011**, *54*, 163–169. [CrossRef]

12. D'Asaro, F.; Grillone, G. Empirical investigation of curve number method parameters in the Mediterranean area. *J. Hydrol. Eng.* **2012**, *17*, 1141–1152. [CrossRef]

13. D'Asaro, F.; Grillone, G.; Hawkins, R.H. Curve number: Empirical evaluation and comparison with curve number handbook tables in Sicily. *J. Hydrol. Eng.* **2014**, *19*, 04014035. [CrossRef]

14. Singh, P.K.; Mishra, S.K.; Berndtsson, R.; Jain, M.K.; Pandey, R.P. Development of a modified SMA based MSCS-CN model for runoff estimation. *Water Res. Manag.* **2015**, *29*, 4111–4127. [CrossRef]

15. Ajmal, M.; Kim, T.-W. Quantifying excess stormwater using SCS-CN–based rainfall runoff models and different curve number determination methods. *J. Irrig. Drain. Eng.* **2015**, *141*, 04014058. [CrossRef]

16. Baltas, E.; Dervos, N.; Mimikou, M. Determination of the SCS initial abstraction ratio in an experimental watershed in Greece. *Hydrol. Earth Sys. Sci.* **2007**, *11*, 1825–1829. [CrossRef]

17. Sharpley, A.; Williams, J. *Epic—Erosion/Productivity Impact Calculator: I. Model Documentation. II: User Manual*; Technical Bulletin, No. 1768 1990; United State Department of Agriculture: Washington, DC, USA, 1990.

18. Williams, J.R.; Izaurralde, R.C.; Steglich, E.M. Agricultural policy/environmental extender model. *Theor. Doc. Version* **2008**, *604*, 2008–2017.

19. Huang, M.; Gallichand, J.; Wang, Z.; Goulet, M. A modification to the soil conservation service curve number method for steep slopes in the Loess Plateau of China. *Hydrol. Process.* **2006**, *20*, 579–589. [CrossRef]

20. Deshmukh, D.S.; Chaube, U.C.; Hailu, A.E.; Gudeta, D.A.; Kassa, M.T. Estimation and comparision of curve numbers based on dynamic land use land cover change, observed rainfall-runoff data and land slope. *J. Hydrol.* **2013**, *492*, 89–101. [CrossRef]

21. Ebrahimian, M.; Nuruddin, A.A.B.; Soom, M.A.B.M.; Sood, A.M.; Neng, L.J. Runoff estimation in steep slope watershed with standard and slope-adjusted curve number methods. *Pol. J. Environ. Stud.* **2012**, *21*, 1191–1202.

22. Mishra, S.K.; Chaudhary, A.; Shrestha, R.K.; Pandey, A.; Lal, M. Experimental verification of the effect of slope and land use on scs runoff curve number. *Water Res. Manag.* **2014**, *28*, 3407–3416. [CrossRef]

23. Williams, J.; Izaurralde, R. The APEX model. In *Watershed Models*; CRC Press: Boca Raton, FL, USA, 2005.

24. Williams, J.; Kannan, N.; Wang, X.; Santhi, C.; Arnold, J. Evolution of the SCS runoff curve number method and its application to continuous runoff simulation. *J. Hydrol. Eng.* **2012**, *17*, 1221–1229. [CrossRef]

25. Fang, H.; Cai, Q.; Chen, H.; Li, Q. Effect of rainfall regime and slope on runoff in a gullied loess region on the Loess Plateau in China. *Environ. Manag.* **2008**, *42*, 402–411. [CrossRef] [PubMed]

26. Garg, V.; Nikam, B.R.; Thakur, P.K.; Aggarwal, S. Assessment of the effect of slope on runoff potential of a watershed using NRCS-CN method. *Int. J. Hydrol. Sci.* **2013**, *3*, 141–159. [CrossRef]

27. Slattery, M.C.; Bryan, R.B. Hydraulic conditions for rill incision under simulated rainfall: A laboratory experiment. *Earth Surf. Process. Landf.* **1992**, *17*, 127–146. [CrossRef]

28. Govers, G. A field study on topographical and topsoil effects on runoff generation. *Catena* **1991**, *18*, 91–111. [CrossRef]

29. Fox, D.; Bryan, R.; Price, A. The influence of slope angle on final infiltration rate for interrill conditions. *Geoderma* **1997**, *80*, 181–194. [CrossRef]

30. Van de Giesen, N.; Stomph, T.J.; de Ridder, N. Surface runoff scale effects in west African watersheds: Modeling and management options. *Agric. Water Manag.* **2005**, *72*, 109–130. [CrossRef]

31. Joel, A.; Messing, I.; Seguel, O.; Casanova, M. Measurement of surface water runoff from plots of two different sizes. *Hydrol. Process.* **2002**, *16*, 1467–1478. [CrossRef]

32. Stomph, T.; De Ridder, N.; Steenhuis, T.; Van de Giesen, N. Scale effects of hortonian overland flow and rainfall-runoff dynamics: Laboratory validation of a process-based model. *Earth Surf. Process. Landf.* **2002**, *27*, 847–855. [CrossRef]

33. Esteves, M.; Lapetite, J.M. A multi-scale approach of runoff generation in a sahelian gully catchment: A case study in Niger. *Catena* **2003**, *50*, 255–271. [CrossRef]

34. Lal, M.; Mishra, S.; Pandey, A.; Pandey, R.; Meena, P.; Chaudhary, A.; Jha, R.K.; Shreevastava, A.K.; Kumar, Y. Evaluation of the soil conservation service curve number methodology using data from agricultural plots. *Hydrogeol. J.* **2017**, *25*, 151–167. [CrossRef]

35. Mah, M.; Douglas, L.; Ringrose-Voase, A. Effects of crust development and surface slope on erosion by rainfall. *Soil Sci.* **1992**, *154*, 37–43. [CrossRef]

36. Kowalik, T.; Walega, A. Estimation of CN parameter for small agricultural watersheds using asymptotic functions. *Water* **2015**, *7*, 939–955. [CrossRef]

37. Lian, H.; Yen, H.; Huang, C.; Feng, Q.; Qin, L.; Bashir, M.A.; Wu, S.; Zhu, A.-X.; Luo, J.; Di, H. CN-China: Revised runoff curve number by using rainfall-runoff events data in China. *Water Res.* **2020**, *177*, 115767. [CrossRef] [PubMed]

38. Qiu, L.; Im, E.-S.; Hur, J.; Shim, K.-M. Added value of very high resolution climate simulations over South Korea using WRF modeling system. *Clim. Dyn.* **2019**, *54*, 173–189. [CrossRef]

39. Isik, S.; Kalin, L.; Schoonover, J.E.; Srivastava, P.; Lockaby, B.G. Modeling effects of changing land use/cover on daily streamflow: An artificial neural network and curve number based hybrid approach. *J. Hydrol.* **2013**, *485*, 103–112. [CrossRef]

40. Mishra, S.K.; Jain, M.K.; Suresh Babu, P.; Venugopal, K.; Kaliappan, S. Comparison of AMC-dependent CN-conversion formulae. *Water Res. Manag.* **2008**, *22*, 1409–1420. [CrossRef]

41. McCutcheon, S.; Tedela, N.; Adams, M.; Swank, W.; Campbell, J.; Hawkins, R.; Dye, C. *Rainfall-Runoff Relationships for Selected Eastern us Forested Mountain Watersheds: Testing of the Curve Number Method for Flood Analysis*; West Virginia Division of Forestry: Charleston, WV, USA, 2006.

42. NRCS (Natural Resources Conservation Service). *National Engineering Handbook Section-4, Part 630, Hydrology*; NRCS: Washington, DC, USA, 2001.

43. Lal, M.; Mishra, S.K.; Pandey, A. Physical verification of the effect of land features and antecedent moisture on runoff curve number. *Catena* **2015**, *133*, 318–327. [CrossRef]

44. Moriasi, D.N.; Arnold, J.G.; Van Liew, M.W.; Bingner, R.L.; Harmel, R.D.; Veith, T.L. Model evaluation guidelines for systematic quantification of accuracy in watershed simulations. *Trans. ASABE* **2007**, *50*, 885–900. [CrossRef]

45. Ritter, A.; Muñoz-Carpena, R. Performance evaluation of hydrological models: Statistical significance for reducing subjectivity in goodness-of-fit assessments. *J. Hydrol.* **2013**, *480*, 33–45. [CrossRef]

46. Harmel, R.; Smith, P.; Migliaccio, K.; Chaubey, I.; Douglas-Mankin, K.; Benham, B.; Shukla, S.; Muñoz-Carpena, R.; Robson, B.J. Evaluating, interpreting, and communicating performance of hydrologic/water quality models considering intended use: A review and recommendations. *Environ. Model. Softw.* **2014**, *57*, 40–51. [CrossRef]

47. Van Mullem, J. Runoff and peak discharges using Green-Ampt infiltration model. *J. Hydraul. Eng.* **1991**, *117*, 354–370. [CrossRef]

48. King, K.W.; Arnold, J.; Bingner, R. Comparison of Green-Ampt and curve number methods on Goodwin Creek watershed using SWAT. *Trans. ASAE* **1999**, *42*, 919. [CrossRef]

49. Walega, A.; Salata, T. Influence of land cover data sources on estimation of direct runoff according to SCS-CN and modified SME methods. *Catena* **2019**, *172*, 232–242. [CrossRef]

50. Hawkins Richard, H. Curve Number Method: Time to Think Anew? *J. Hydrol. Eng.* **2014**, *19*, 1059. [CrossRef]

51. Soulis, K.X.; Valiantzas, J.D. Identification of the SCS-CN parameter spatial distribution using rainfall-runoff data in heterogeneous watersheds. *Water Res. Manag.* **2013**, *27*, 1737–1749. [CrossRef]

52. Soulis, K.X. Estimation of SCS curve number variation following forest fires. *Hydrol. Sci. J.* **2018**, *63*, 1332–1346. [CrossRef]

Article

An Improved SCS-CN Method Incorporating Slope, Soil Moisture, and Storm Duration Factors for Runoff Prediction

Wenhai Shi [1,2,*] and Ni Wang [1,2]

[1]   Key Laboratory of Subsurface Hydrology and Ecological Effects in Arid Region, Chang'an University, Ministry of Education, Xi'an 710054, China; Wangnivviv@chd.edu.cn

[2]   School of Water and Environment, Chang'an University, Xi'an 710054, China

*   Correspondence: shiwenhai@chd.edu.cn; Tel.: +86-29-82339291; Fax: +86-29-82339291

Received: 9 April 2020; Accepted: 7 May 2020; Published: 8 May 2020

Abstract: Soil Conservation Service Curve Number (SCS-CN) is a popular surface runoff prediction method because it is simple in principle, convenient in application, and easy to accept. However, the method still has several limitations, such as lack of a land slope factor, discounting the storm duration, and the absence of guidance on antecedent moisture conditions. In this study, an equation was developed to improve the SCS-CN method by combining the CN value with the tabulated CN_2 value and three introduced factors (slope gradient, soil moisture, and storm duration). The proposed method was tested for calibration and validation with a dataset from three runoff plots in a watershed of the Loess Plateau. The results showed the model efficiencies of the proposed method were improved to 80.58% and 80.44% during the calibration and validation period, respectively, which was better than the standard SCS-CN and the other two modified SCS-CN methods where only a single factor of soil moisture or slope gradient was considered, respectively. Using the parameters calibrated and validated by dataset of the initial three runoff plots, the proposed method was then applied to runoff estimation of the remaining three runoff plots in another watershed. The proposed method reduced the root-mean-square error between the observed and estimated runoff values from 5.53 to 2.01 mm. Furthermore, the parameters of soil moisture (b_1 and b_2) is the most sensitive, followed by parameters in storm duration (c) and slope equations (a_1 and a_2), and the least sensitive parameter is the initial abstraction ratio λ on the basis of the proposed method sensitivity analysis. Conclusions can be drawn from the above results that the proposed method incorporating the three factors in the SCS method may estimate runoff more accurately in the Loess Plateau of China.

Keywords: storm duration; soil moisture; slope; Soil Conservation Service Curve Number method; runoff prediction

1. Introduction

The ability to make surface runoff estimations has become an essential part of development strategies in water resources management, flood control, and water and soil conservation [1]. A multitude of hydrologic models have been developed to predict direct runoff. Among these methods, the Soil Conservation Service Curve Number (SCS-CN) method [2] is one of the most widely used. The method was originally developed for surface runoff prediction, but has been applied to several other areas such as the infiltration, sediment yield, and transport of pollutants [3,4]. Moreover, it has been extensively integrated into many hydrological and ecological models [5], including AnnAGNPS (Annualized Agricultural Nonpoint Source Pollution Model) [6], CREAMS (Chemicals, Runoff, and Erosion from Agricultural Management Systems) [7], EPIC (Environmental

Policy Integrated Climate) [8], SWAT (Soil and Water Assessment Tool) [9], and EBA4SUB (Event-Based Approach for Small and Ungauged Basins) [10,11].

The popularity of the SCS-CN method is due to its simplicity, convenience, widespread acceptance, applicability to ungauged watersheds, and requirement of only one parameter (*CN*), which is determined by four readily grasped watershed characteristics (soil group, land cover, surface condition, and antecedent moisture condition (AMC)) [12,13]. However, the SCS-CN also has some limitations such as lack of a land slope factor, discounting the storm duration, and the absence of guidance on antecedent moisture conditions [14–19]. These advantages and disadvantages made the SCS-CN method a perennial topic of discussion over the last four decades [20–25].

Storm duration is a vital part of the rainfall-runoff process, which could greatly influence runoff estimations [14]. However, the storm duration is not considered in the SCS-CN method, which leads to uncertainty in runoff prediction because of the spatiotemporal variability of rainfall [12]. Several enhanced SCS-CN methods have been proposed to overcome the problem of discounting the storm duration. Jain et al. [17] suggested a new model formulation with storm duration-based rainfall adjustment; Mishra et al. [18] presented a rainfall duration-dependent procedure by developing an equation for the new *CN* value with an introduced minimum *CN* and rainfall duration; Petroselli et al. [26] presented a model for continuous simulation and the SCS-CN method was modified to calculate the net rainfall by resolving the problem of insufficient including infiltration; Sahu et al. [27] incorporated the storm duration factor into the SCS-CN method by dividing the cumulative infiltration amount *F* into dynamic infiltration and static infiltration (F_c), which is a product of the minimum infiltration rate and rainfall duration; Shi et al. [25] introduced the static infiltration into the soil moisture accounting (SMA)-based SCS-CN method to improve runoff predictions on the Loess Plateau. However, all these methods have no contact with the *CN* value of the original SCS-CN method, which has limited the application of the models.

Antecedent moisture condition (AMC) is a significant factor determining the initial abstraction of the SCS-CN method. The AMC prior to a storm event is divided into three levels: dry (AMC 1), average (AMC 2), and wet (AMC 3) based on the five-day antecedent rainfall amount in the traditional SCS-CN method, which will cause the *CN* value to suddenly jump from one level to another [28]. Soil moisture seems to be a better choice for selecting an appropriate AMC value as compared with the five-day antecedent rainfall depth, and improved knowledge of the relationship between antecedent moisture and the *CN* value would greatly improve the runoff estimation of the SCS-CN method [29,30]. Saxton [31] redefined the antecedent moisture condition with a step function of soil water content. Koelliker [32] suggested that *CN* changes linearly with the soil moisture. The EPIC and SWAT models also adopted a continuous equation between soil moisture and *CN* for runoff prediction. Jacobs et al. [33] used remote sensing of soil moisture for *CN* adjustment to predict runoff in five basins of Oklahoma. Based on the climatic conditions of the Loess Plateau in China, Huang et al. [34] established an equation between soil water content and *CN* value for better runoff estimation.

The land slope is another main factor affecting flow movement within the landscape. The *CN* values in the USDA-NRCS (US Department of Agriculture–Natural Resources and Conservation Service) handbook table are based on 5% slope, and should be adjusted according to the actual slope. Several efforts have also been made on consideration of a slope factor in the CN method. Sharpley and Williams [8] introduced a slope-adjusted CN_2 in the calculation of the runoff volume, but it has not been tested in the field. Based on the data of experimental plots with slopes varying from 14% to 140% on the Loess Plateau of China, Huang et al. [15] proposed an equation for considering the influence of slope on *CN* value. However, there is no study that couples these factors (storm duration, soil moisture, and slope) in the SCS-CN method for predicting runoff, which would be the focus of this study.

Therefore, the objectives of this study are (1) to develop an equation of the *CN* value with original CN_2 value, slope gradient, soil moisture, and storm duration in the conventional SCS-CN method; (2) to compare the performance of the standard SCS-CN method, those from Huang et al. [15] and Huang et al. [34], and the proposed method by observing three experimental plots in the Loess

Plateau region; (3) to apply the proposed method to predict runoff from three runoff plots in the other watershed.

2. Methods

2.1. The Original SCS-CN Method

The SCS-CN method [2] was originally developed based on two fundamental hypotheses and one simple water balance equation, which can be written as follows:

$$P = I_a + F + Q \tag{1}$$

$$\frac{Q}{P - I_a} = \frac{F}{S} \tag{2}$$

$$I_a = \lambda S \tag{3}$$

where P and Q are the depth of observed rainfall and direct runoff, respectively (mm); I_a and λ are the initial abstraction (mm) and coefficient of initial abstraction (dimensionless), respectively; F is the cumulative amount of infiltration (mm); and S is the maximum potential retention (mm), which can be calculated by

$$S = \frac{25400}{CN} - 254 \tag{4}$$

where CN is in the range of 0–100 (dimensionless). This CN value is determined by the CN value (CN_2) of the average moisture condition (AMC 2), which depends on land cover, soil group, and hydrologic conditions using a table from the SCS handbook [2], and is then converted to AMC 1 or AMC 3 based on the five-day-prior precipitation.

2.2. The Proposed Method

In the proposed method, slope gradient, soil moisture, and storm duration factors are considered. The CN value for each rainfall-runoff event can be improved by multiplying the reference value of SCS handbook CN_2, and the function of slope, soil moisture, and storm duration:

$$CN = CN_2 f(\alpha) f(\theta) f(t) \tag{5}$$

where α is the slope gradient (m m^{-1}), θ is soil moisture (cm^3 cm^{-3}), and t is storm duration (h). $f(\alpha)$, $f(\theta)$, and $f(t)$ are functions of α, θ, and t, respectively.

The slope equation proposed by Huang et al. [15] was used to consider the influence of slope on the CN_2 value. They used the modified SCS-CN method to evaluate runoff prediction in runoff plots with an 11-year observation experiment and a slope range of 14–140% in Xifeng City of the Loess Plateau. It is adopted in this study:

$$f(\alpha) = \frac{a_1 + a_2(\alpha - 0.05)}{(\alpha - 0.05) + a_1} \tag{6}$$

A nonlinear equation between CN and soil moisture was developed by Huang et al. [34], who found that the V_0 value prior to each storm event can be expressed by the water storage in the upper 15 cm soil. It is also adopted and can be expressed as follows:

$$f(\theta) = \frac{\theta}{b_1 + b_2\theta} \tag{7}$$

For the storm duration function, we proposed a one-parameter (c) equation for $f(t)$ after several trials:

$$f(t) = 1 - ct \tag{8}$$

where a_1, a_2, b_1, b_2, and c are the empirical coefficients.

Therefore, Equations (5)–(8) combined with Equations (1)–(4) make up the proposed method for runoff prediction. The flowchart of the whole methodology applied in this study is shown in Figure 1.

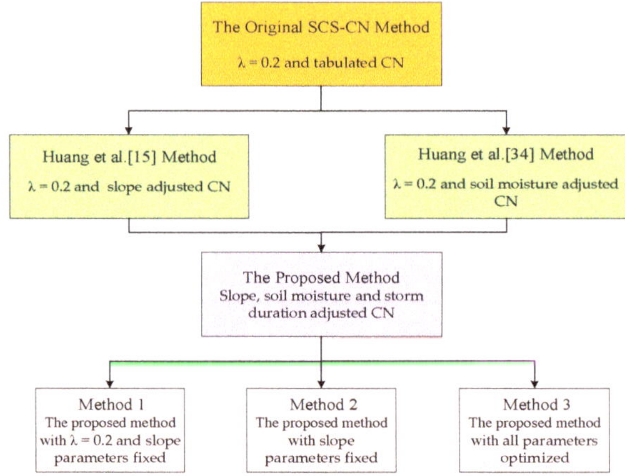

Figure 1. The flowchart of the whole methodology applied in this study.

2.3. Performance of the Methods

The following two statistical indices were used to evaluate the performance of the methods:

$$NSE = \left[1 - \frac{\sum\limits_{i=1}^{N}\left(Q_i - Q_i^*\right)^2}{\sum\limits_{i=1}^{N}\left(Q_i - \overline{Q_i}\right)^2}\right] \times 100\% \tag{9}$$

$$RMSE = \sqrt{\frac{1}{N}\sum\limits_{i=1}^{N}\left(Q_i - Q_i^*\right)^2} \tag{10}$$

where NSE is the Nash–Sutcliffe efficiency [35,36] and $RMSE$ is the root-mean-square error; Q_i and Q_i^* are the ith observed and estimated runoff, respectively; $\overline{Q}$ is the average measured runoff of events; and N is the total events. Higher NSE and lower $RMSE$ values indicate that the model has better agreement with the observations.

3. Study Area and Data

3.1. Study Area

This study was conducted in six test plots, three each in the Xindiangou (XDG) and Chabagou (CBG) watersheds, which are both in the hilly region of the Loess Plateau (Figure 2).

The XDG watershed (latitude: 37°29′ N, longitude: 110°08′ E, elevation: 990–1010 MASL, area: 30 km²) is located in Suide County. The average annual temperature is 8 °C and the mean annual precipitation is 470 mm, mostly between June and September. The soil type of the XDG watershed is mainly silty loam. The characteristics of soil physical and particle size distribution are homogeneous in the top 30 cm soil [37].

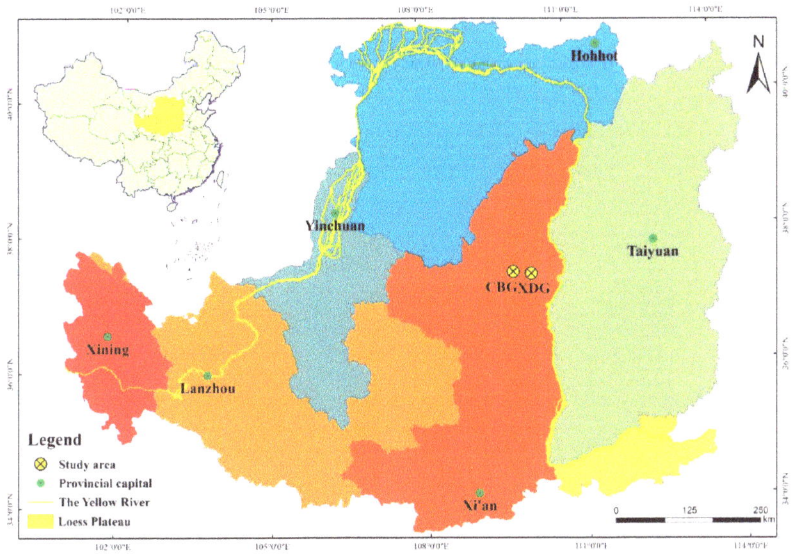

Figure 2. Location of the two experimental watersheds of Chabagou (CBG) and Xindiangou (XDG).

The CBG watershed (latitude: 37°31′ N, longitude: 109°47′ E; elevation: 900–1100 MASL; area: 205 km^2) is located in Zizhou County. The mean temperature is also 8 °C and the average annual precipitation is 450 mm. The precipitation from July to August accounts for more than 70%, most of which are heavy and short-term rainstorms. The main soil in CBG watershed is Malan loess soil, in which the content of clay particles is less than 40%, leading to the large porosity of soil and the vulnerability to erosion [38].

In the XDG watershed, alfalfa (*Medicago sativa* L.), sweet clover (*Melilotus suavcolcn*), and millet (*Setaria italica*) were planted in the three runoff plots during the observation period from 1954 to 1959. In the CBG watershed, the three experimental plots were grown with pasture (*Arundinella hirta*), potatoes (*Solanum tuberosum* L.), and millet during the test period from 1959 to 1965, respectively (Table 1).

Table 1. Main characteristics of experimental plots in the XDG and CBG watersheds.

Watershed	XDG			CBG		
Plot	1	2	3	1	2	3
Land use	Grassland	Grassland	Cropland	Grassland	Cropland	Cropland
Vegetation	Alfalfa	Sweet clover	Millet	Pasture	Millet	Potato
Length (m)	20	20	20	40	20	20
Width (m)	5	5	5	10	10	10
Slope gradient (°)	35	33	15	30	25	22
Soil moisture (%)	18.54 ± 5.93	18.08 ± 5.92	19.68 ± 5.89	15.82 ± 4.50	16.80 ± 5.64	16.79 ± 4.30
Rainfall (mm)	23.59 ± 14.43	23.90 ± 22.76	25.97 ± 22.08	22.31 ± 24.31	24.81 ± 28.24	23.93 ± 17.88
Runoff (mm)	4.52 ± 8.54 [#]	4.60 ± 7.97	4.09 ± 4.52	2.07 ± 2.12	2.02 ± 1.95	6.61 ± 8.16
Strom duration (h)	4.53 ± 6.08	4.88 ± 6.53	5.84 ± 6.95	4.65 ± 6.74	5.82 ± 7.75	2.46 ± 2.59
Observation period	1956–1959	1956–1959	1954–1959	1959–1961	1959–1962	1964–1965

[#]: Mean ± standard deviation.

3.2. Data Collection

Rainfall characteristics in the six plots, including depth and duration of each rainfall, were monitored by self-recording rain gauges. For all the experimental sites in the two watersheds, the rain gauges were located less than 500 m from each experimental plot. At the downstream end of each plot, a funnel-type collector was adopted to collect the surface runoff (Figure 3). Soil samples in each plot were collected at an interval of 10 cm for 0–50 cm soil layer at 10:00 a.m. during the monitoring period, and the gravimetric water content of the collected soil samples was then determined by the oven-drying method. The water content of the top 15 cm soil could be calculated from the mean soil water measurements at 10 and 20 cm soil depth. These datasets have been compiled based on rainfall-runoff events and printed for internal use by the Yellow River Administration Committee of the Ministry of Water Resources and the Shaanxi Institute of Soil and Water Conservation.

Figure 3. The layout of field plots in XDG watershed.

The standard SCS-CN and proposed methods were tested using data derived from the three plots in the XDG watershed for calibration and validation, and the tested methods were then applied to the remaining three runoff plots in the CBG watershed.

3.3. Parameter Estimation

The rainfall-runoff events of the three plots in the XDG watersheds were divided into two groups: the datasets of cropland and one grassland plot were used for calibration, and the remaining datasets of another grassland plot were used for validation. The Marquardt [39] algorithm for solving constrained least squares problems was adopted to optimize all the parameters of the test methods. In order to examine the applicability of the models, $\lambda = 0.2$, assigned in its original development, is adopted for the traditional SCS-CN, Huang et al. [15], Huang et al. [34], and the proposed method (Method 1). The slope parameters (a_1 and a_2) obtained from Huang et al. [15] were also used for Huang et al. [15] and the proposed method (Method 1). Moreover, the effect of initial abstraction and slope on the proposed method was tested with the optimized λ (Method 2) and both optimized λ and slope parameters (Method 3).

4. Results

4.1. Model Calibration and Validation

The number of storm events observed for calibration and validation in watershed XDG was 47 and 29, respectively. Table 2 presents the parameters of the tested methods optimized with the calibration dataset. The overall performance of all tested models based on statistical indicators is compared in Table 3.

Table 2. The optimized parameters of the tested methods for the plots in XDG watershed.

Model	Parameter					
	λ	a_1	a_2	b_1	b_2	c
Original SCS-CN	0.2	-	-	-	-	-
Huang et al. [15]	0.2	323.57	15.63	-	-	-
Huang et al. [34]	0.2	-	-	0.01	1.12	-
Method 1	0.2	323.57	15.63	0.05	0.61	0.020
Method 2	0.001	323.57	15.63	0.13	0.31	0.035
Method 3	0.001	213.99	25.38	0.15	0.24	0.035

Method 1: the proposed method with λ and slope parameters fixed. Method 2: the proposed method with slope parameters fixed. Method 3: the proposed method with all parameters optimized.

Table 3. The performance of the tested method for the calibration and validation datasets.

Model	Linear Regression			NSE	RMSE	Performance
	Slope	Interception	R^2	(%)	(mm)	Rating [40]
Calibration						
Original SCS-CN	0.83	0.07	0.24	−118.64	9.83	Unsatisfactory
Huang et al. [15]	0.92	0.10	0.27	−133.00	10.14	Unsatisfactory
Huang et al. [34]	0.27	0.41	0.15	−9.80	6.96	Unsatisfactory
Method 1	0.88	−0.48	0.79	75.84	3.26	Acceptable
Method 2	0.80	0.95	0.80	80.58	2.94	Good
Method 3	0.82	0.72	0.81	80.73	2.91	Good
Validation						
Original SCS-CN	0.77	1.33	0.18	−182.60	13.18	Unsatisfactory
Huang et al. [15]	0.87	1.58	0.19	−229.67	14.23	Unsatisfactory
Huang et al. [34]	0.21	1.14	0.07	−32.27	9.02	Unsatisfactory
Method 1	0.92	0.04	0.80	77.45	3.72	Acceptable
Method 2	0.81	0.87	0.81	80.44	3.46	Good
Method 3	0.84	0.71	0.81	81.21	3.40	Good
Full data						
Original SCS-CN	0.80	0.56	0.21	−148.17	11.23	Unsatisfactory
Huang et al. [15]	0.90	0.67	0.23	−177.64	11.87	Unsatisfactory
Huang et al. [34]	0.24	0.70	0.11	−20.18	7.81	Unsatisfactory
Method 1	0.90	−0.30	0.79	76.58	3.45	Acceptable
Method 2	0.80	0.92	0.81	80.50	3.15	Good
Method 3	0.83	0.71	0.81	80.95	3.11	Good

R^2: coefficient of determination; *NSE*: model efficiency; *RMSE*: root-mean-square error; Performance rating: very good (NSE > 90%), good (80% < NSE < 90%), acceptable (65% < NSE < 80%), and unsatisfactory (NSE < 65%).

4.1.1. The Original SCS-CN Method

Figure 4 presents the runoff estimated by the original method plotted against the corresponding observed values for the full datasets in XDG watershed. The standard SCS-CN method underpredicted the large and overpredicted as well as underpredicted some small storm-runoff events (Figure 4a). This is also confirmed by the statistical indexes in Table 3. There is a similar pattern in the calibration and validation sub-datasets, where the slopes of regression are 0.83 and 0.77, and the *NSE* values is −118.64% and −182.60%, respectively (Table 3), which illustrates the poor performance of the conventional SCS-CN method.

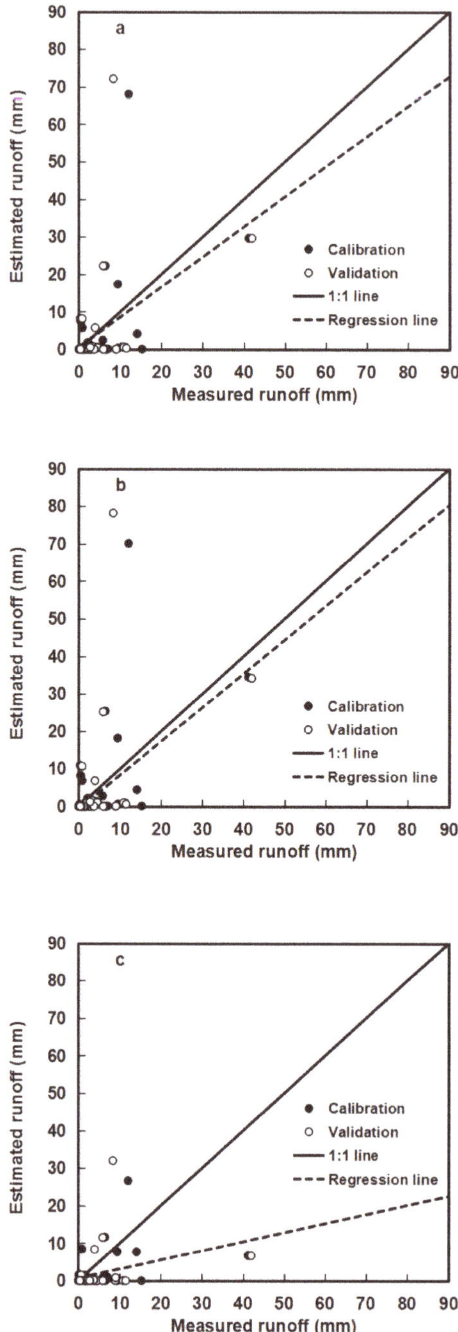

Figure 4. Measured versus estimated runoff depths for (**a**) the original Soil Conservation Service Curve Number (SCS-CN) method, (**b**) Huang et al. [15] method, and (**c**) Huang et al. [34] method for the calibration and validation of the three runoff plots located in the XDG watersheds.

4.1.2. The Huang et al. and Huang et al. Methods

The Huang et al. [15] and Huang et al. [34] methods incorporated the slope gradient and soil moisture condition factors into the standard SCS-CN method. The prediction of large storm-runoff events was improved with a higher slope of 0.90 using the Huang et al. [15] method as compared with the conventional SCS-CN method (0.80) (Figure 4b). However, it also overpredicted some small storm-runoff events when the slope factor is considered in the original SCS-CN method. Therefore, it can be seen that the Huang et al. [15] method did not improve the prediction accuracy of the full dataset much, as verified by the *NSE* value, which decreased from −148.17% (original SCS-CN method) to −177.63% (Table 3)

Table 3 also shows that the *NSE* value (−9.8%) of the Huang et al. [34] method during the calibration period was significantly improved as compared to the SCS-CN (−118.64%) and Huang et al. [15] method (−133.00%). The overprediction of the latter two methods incurred for some small storm-runoff events were mitigated. However, the regression slope of the Huang et al. [34] method deviates more from the 1:1 line more than the traditional SCS-CN and Huang et al. [15] method, which suggests that the former still underestimated large and overestimated as well as underestimated small runoff events (Figure 4c).

4.1.3. The Proposed Method (Methods 1–3)

The runoff estimated by Method 1 is closer to the observed values, with the coefficient of determination R^2 increasing to 0.79 from 0.21 (the SCS-CN method), and the regression line of which is closer to the perfect line, where the regression line slope and intercept is 0.90 and 0.30, respectively (Table 3). Moreover, the prediction results of Method 1 are satisfactory because most of the values are closer to the perfect prediction line, and in good agreement with observed runoff even for large runoff events (Figure 5a). Method 1 yielded a larger *NSE* value of 75.84% and a smaller *RMSE* value of 3.26 mm in the calibration case, whereas the *NSE* value of Method 1 increases from −182.60% (the traditional SCS-CN method) to 77.45%, and the *RMSE* value decreased from 13.18 to 3.72 mm in the validation case (Table 3). Thus, we can conclude that Method 1 performed the best of the four methods for both the calibration and validation cases.

However, the points of Method 1 around the 1:1 line are still scattered, which might be inherent in the CN method where the initial abstraction ratio λ is set to a standard value of 0.2. The assumption that $\lambda = 0.2$ is often questioned by many scholars [12,41]. Therefore, in order to improve the SCS-CN method, we introduced Method 2 with an optimized λ based on Method 1 to test the effect of initial abstraction on the proposed method. Moreover, we also optimized the parameters in Equation (6) (a_1 and a_2) (Method 3) based on Method 2 to test the slope factor on the proposed method as compared with the fixed values obtained from Huang et al. [15].

Figure 5b,c presents the comparison of runoff predicted by Methods 2 and 3 and the corresponding observed values, respectively. It can be seen that the runoff estimation is a great improvement over Method 1, with observation points close to the perfect line, especially when the depths of runoff are between 10 and 20 mm. The statistics also demonstrate the predictive capacity of Methods 2 and 3 with a higher model efficiency of 80.50% and 80.95%, respectively, in Table 3. However, there is a slight improvement with Method 3, which uses an optimized slope parameter ($a_1 = 213.99$ and $a_2 = 25.38$), as compared with Method 2, with *NSE* values of 80.73% vs. 80.44% in calibration and 81.21% vs. 80.58% in validation. The results indicated that the proposed methods with optimized λ (Methods 2 and 3) could further improve the SCS-CN method for runoff prediction in this study area.

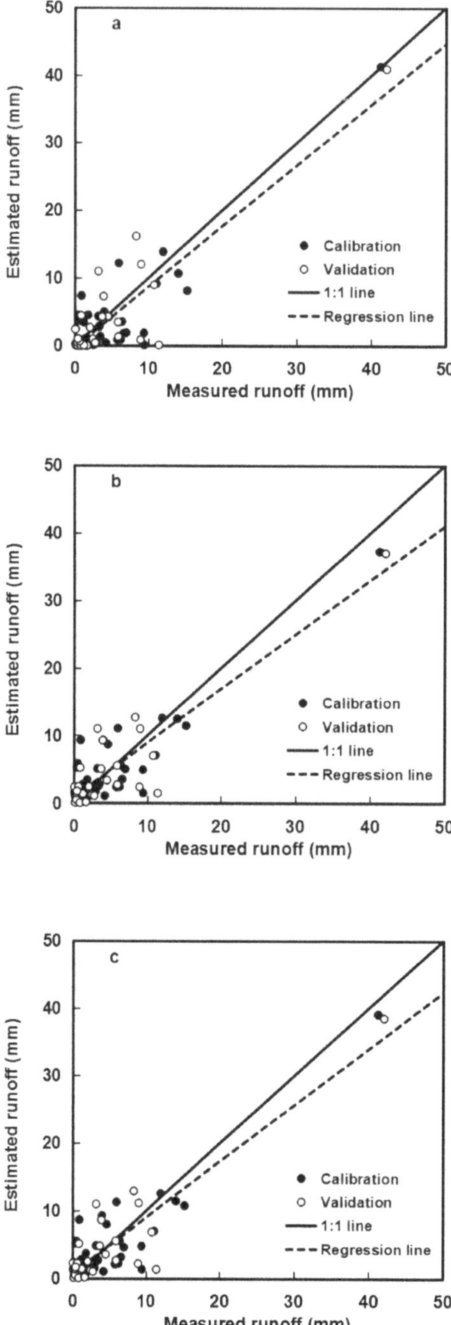

Figure 5. Measured versus estimated runoff depths for (**a**) Method 1, (**b**) Method 2, and (**c**) Method 3 for calibration and validation for three runoff plots of the XDG watersheds.

4.2. Model Application

The tested standard SCS-CN and the proposed method (Methods 1–3) using the parameters calibrated and validated by the dataset of the three plots in XDG watershed were applied to estimate runoff of the remaining three plots in the CBG watershed.

Figure 6 shows the measured versus estimated runoff values for the conventional SCS-CN and proposed method with the dataset in plots of the CBG watershed. The original SCS-CN method continued to underpredict most of the large as well as small storm-runoff events, whereas Method 1 performed better than the original SCS-CN method, with the former yielding a higher *NSE* (75.33%) and a smaller *RMSE* value (2.38 mm) (Figure 6). However, Method 1 with λ = 0.2 still underestimated some small storm-runoff events as compared with Method 2 (λ = 0.001), which achieved a higher *NSE* value of 82.29% and had most of its data points visibly closer to the perfect line. Moreover, Method 2, using the parameters obtained from Huang et al. [15], with data from slopes ranging from 14% to 140%, performed better than Method 3 (*NSE* = 79.22%), which used optimized parameters with a narrow range of slope (26.7% and 70%). Therefore, from the results of the experimental sites, Method 2 is the most suitable SCS-CN method for surface runoff estimation in the plots of the two watersheds, which may also be applicable in the Loess Plateau.

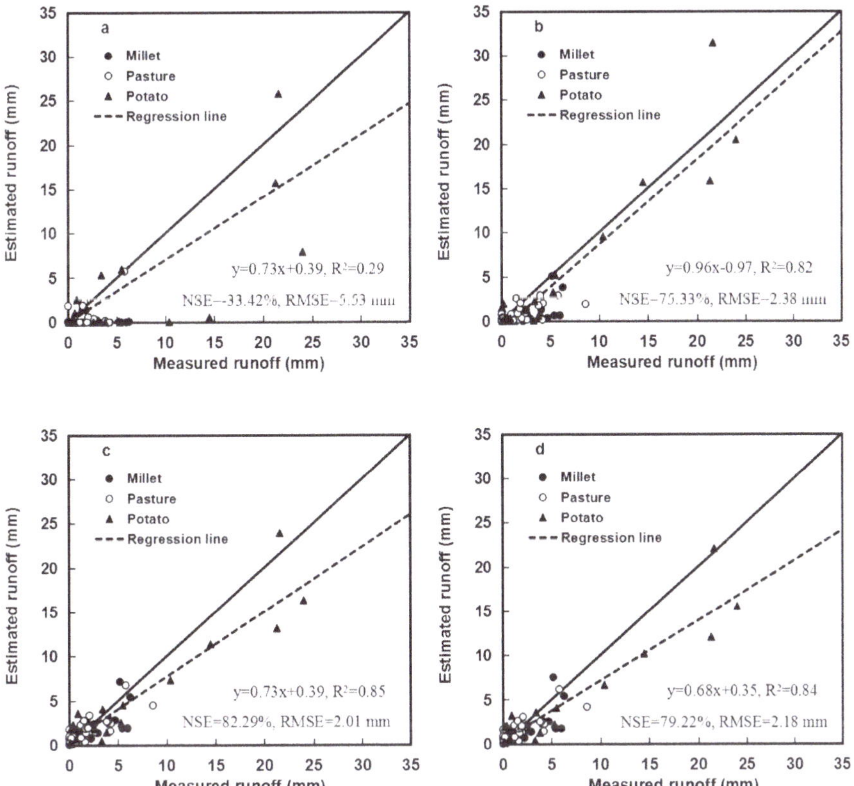

Figure 6. Measured versus estimated runoff depths for (**a**) original SCS-CN method, (**b**) Method 1, (**c**) Method 2, and (**d**) Method 3 for the three experimental plots located in the CBG watershed.

4.3. Sensitivity Analyses

The above results indicate that Method 2 predicts runoff with greater accuracy than the other SCS-CN methods. A sensitivity analysis can identify the primary parameters that affect the performance of the model. In this study, sensitivity analyses were conducted by observing the effect of a variation in the calibrated parameter on the runoff prediction in terms of *NSE*, using the datasets of the plots in XDG watershed.

Figure 7 presents the sensitivity of the runoff estimations to each parameter of Method 2. A sensitive variable is a parameter for which *NSE* changes dramatically as the parameter increases or decreases from the calibrated value. The most sensitive variable is apparently parameter b_1, where the value of *NSE* sharply decreases from 80.58% to 47.83% as the b_1 value changes in a small range of 130–80% of the calibrated value. However, the initial ratio λ appears to be the least sensitive. In general, the parameters of soil moisture (b_1 and b_2) are the most sensitive, followed by parameters in storm duration (c) and slope equations (a_1 and a_2), and the least sensitive parameter is the initial abstraction ratio λ.

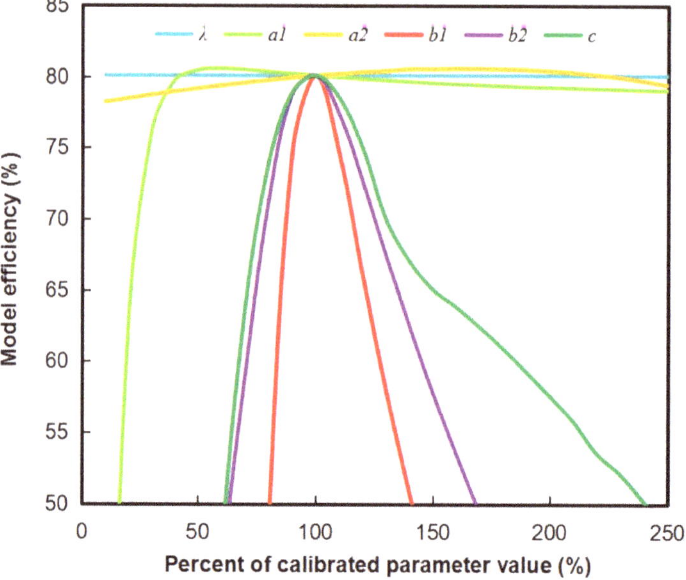

Figure 7. Sensitivity analysis of the empirical parameters of Method 2.

5. Discussion

From the above results, the original SCS-CN method had no good performance for the three plots of the XDG watersheds in both calibration and validation cases; it still underpredicted numerous rainfall-runoff events, and it also performed poorly when applied to the plots of the CBG watershed. The poor quality of the SCS-CN method might be due to the *CN* value that was taken from the USDA tables [42], which need improvement either by accounting for different factors that influence runoff or rebuilding the *CN* values based on different climate, soil, land use, and slope conditions from a large set of monitored rainfall-runoff data. In this study, several factors that influence runoff are considered to improve the *CN* value. Although the Huang et al. [15] and Huang et al. [34] methods, in which single factors of slope gradient and soil moisture were considered, respectively, could enhance the performance to some extent, the improvement is limited. Overall, the above three methods did not perform as well as the proposed method. The better performance of the proposed method shows it can accurately estimate runoff in the six plots with different vegetation types in the XDG and CBG watersheds using the optimized parameters. The results are also given some understanding of the physical processes represented by incorporating the slope, soil moisture, and storm duration factors into the proposed method, which eliminated the main drawbacks of the conventional SCS-CN model.

5.1. The Effect of Rainfall Duration

The characteristics of rainfall are the main driving force of runoff process, and different rainfall regimes will lead to different capacities of runoff production for each plot.

The 148 rainfall events of the six plots in the two watersheds (XDG and CBG) were divided into three categories using K-means clustering [43] based on rainfall depth and duration to test the performance of the SCS-CN method for different types of rainfall regimes (Table 4). The classification conformed to the ANOVA criterion of a significant level (*** $p < 0.001$). Rainfall regime 1 is a group of storm events with heavy intensity, short rainfall duration, and lower precipitation, and its occurrence frequency is the highest, accounting for 79.73% of the total events. Rainfall Regime 3 consists of storm events with low intensity, long-term duration, and low frequency (3.38%), while Rainfall Regime 2 is composed of storm events with moderate storm characteristics, i.e., shorter rainfall duration and lower precipitation than Rainfall Regime 3, but longer rainfall duration and higher precipitation than Rainfall Regime 1. In general, the descending order of average P and D is as follows: Rainfall Regime 3 > Rainfall Regime 2 > Rainfall Regime 1. This classification of rainfall regimes is consistent with Wei et al. [44] and Fang et al. [45], who also classified the rainfall into three rainfall regimes with 14 and 9 years of field measurements on the Loess Plateau, respectively.

Table 4. Statistical features of the rainfall regimes in XDG and CBG watersheds.

Rainfall Regime	Eigenvalue	Mean	Standard Deviation	Variation Coefficient	Frequency (%)
Regime 1	P (mm)	14.92	6.95	0.47	79.73
	D (h)	2.43	2.55	1.05	
Regime 2	P (mm)	50.49	11.06	0.22	16.89
	D (h)	12.61	7.94	0.63	
Regime 3	P (mm)	109.36	7.68	0.07	3.38
	D (h)	22.93	1.31	0.06	

P: precipitation depth; D: storm duration.

Figure 8 shows the relationship between measured and estimated runoff values using the traditional SCS-CN and Method 2 under the three rainfall regimes. The original SCS-CN method underpredicted most storm-runoff events of the Rainfall Regime 1 and overpredicted the events of Rainfall Regime 3, while Rainfall Regime 2 performed better than Rainfall Regimes 1 and 3, but the points of Rainfall Regime 2 are still scattered around the 1:1 line (Figure 8a). This is because the SCS-CN method accounts for the rainfall amount but ignores the storm duration. The runoff predicted by the conventional SCS-CN method increased as the rainfall amount increased from Rainfall Regimes 1 to 3, while the storm duration also increased from Rainfall Regimes 1 to 3. Therefore, the measured runoff did not increase with the rainfall amount, which ultimately meant that the SCS-CN method performed poorly. However, when incorporating the storm duration in the proposed method with Equation (8), the underprediction of Rainfall Regime 1 and overprediction of Rainfall Regime 3 were removed and the performance of predicting Rainfall Regime 2 was improved, with most of its data points lying close to 1:1 as compared with the traditional SCS-CN method (Figure 8b). The better performance indicated that storm duration plays a vital role in rainfall-runoff generation and estimation, and the proposed method can account for different rainfall regime types of varying duration. Others have also confirmed that storm duration plays a significant role in both runoff production and prediction [45–47].

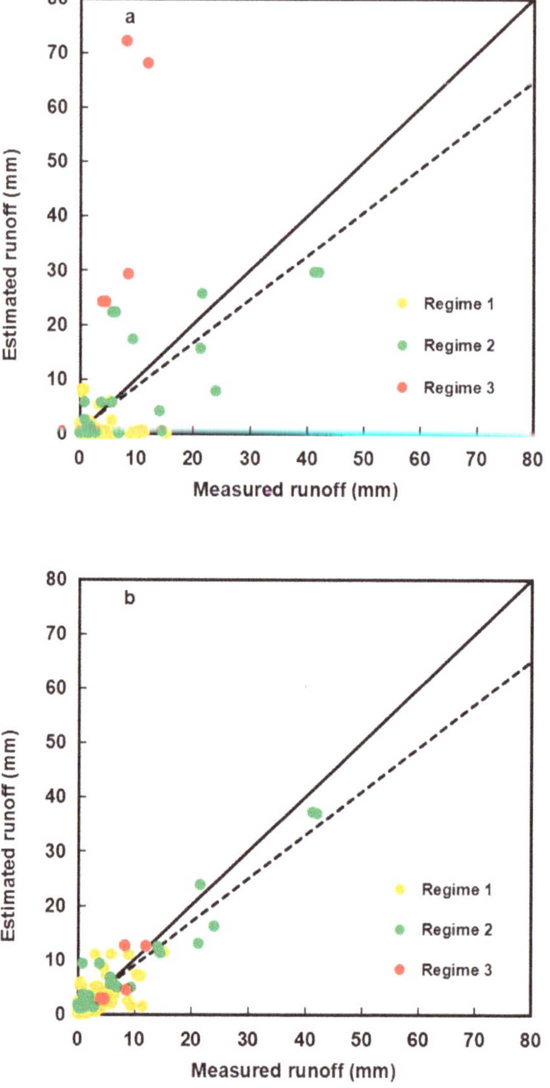

Figure 8. Measured versus estimated runoff depths for (**a**) the original SCS-CN method and (**b**) Method 2 of the three rainfall regimes.

5.2. The Effect of Slope

As the slope increased, the initial abstraction, infiltration, and recession time of overland flow decreased, which resulted in less of an opportunity for infiltration and more surface runoff as compared with a horizontal surface. These drawbacks have been demonstrated by many researchers [48–52]. However, the CN_2 of the standard SCS-CN method obtained from the handbook table [2] are based on 5% slope. Therefore, a slope equation (Equation (6)) was developed that incorporated the slope in the standard SCS-CN method for steep slope conditions.

In this study, the effect of slope is not reflected as the Huang et al. [15] method performed worse than the original SCS-CN method when accounting for the slope factor in the method. The reason is that the traditional SCS-CN has already overpredicted the storm events of Rainfall Regime 3, and the overprediction intensified when the slope factor was taken account in the Huang et al. [15] method because the runoff increased with slope gradient (Figure 4). However, when the overprediction was removed by the proposed method, the effect of slope could be found easily and as the slope increased the runoff increased.

5.3. The Effect of Soil Moisture

The antecedent moisture condition (AMC) plays a significant role in the progress of runoff. The conventional SCS-CN method divides the AMC into three levels based on the five-day antecedent rainfall amount. The measured values of CN should be continuous values for runoff prediction and are not limited to the three discrete AMC levels determined by the five-day AMC, which causes a sudden jump in runoff prediction [25]. Moreover, Koelliker [53] found that the probability of AMC_2 occurring before a rainfall-runoff event was only 7%, rather than 50% [54], which indicated that the AMC value was underestimated and thereby resulted in runoff underprediction by the original SCS method. Huang et al. [34] indicated that there was no correlation between the CN value and the five-day AMC, and the tabulated CN values were less than the measured ones in most cases, which further confirmed the runoff underprediction. Therefore, the five-day AMC used in the original CN method is unreasonable.

An alternative to soil moisture for determining the AMC would be to use direct measured or predicted soil moisture prior to a storm event, instead of antecedent rainfall, which is a vital factor in determining the initial abstraction of the SCS method. The rainfall-runoff prediction will be greatly enhanced by improving knowledge of antecedent moisture [29,30]. The results of the Huang et al. [34] method performed better than Huang et al. [15] and the standard SCS-CN method (Figure 4), which indicated that the equation of soil moisture (Equation (7)) is more suitable for AMC than the antecedent five-day rainfall.

In the new CN value, three factors of slope, soil moisture, and storm event were introduced, which represented the land condition, antecedent moisture, and characteristics of the storm event, respectively; all three factors have a great influence on runoff. However, as compared with the Huang et al. [15] and Huang et al. [34] methods, the performance of Method 1 was significantly improved after accounting for the storm duration factor (Table 3). Thus, storm duration has a greater effect on the runoff prediction of the SCS-CN method, which has frequently been ignored in other modified SCS-CN methods, followed by soil moisture and slope. Moreover, $\lambda = 0.001$ is more appropriate for the initial abstraction coefficient in this study area according to the comparison between Method 1 and 2; the same conclusion was also obtained by Huang et al. [34] for four plots in the Loess Plateau. The slope parameters in Method 2, which is obtained from Huang et al. [15] with data from slopes ranging from 14% to 140%, are also more suitable for application as compared with Method 3. This inference can be further confirmed by the sensitivity analysis, which also suggested that the initial abstraction ratio λ and the slope parameters (a_1 and a_2) are less sensitive, so Method 2 can be applied to similar sub-humid, semi-arid, and arid regions with the optimized parameters, but may need adjustment for humid regions because the soil moisture and rainfall may differ from the tested results in this study.

6. Conclusions

In this study, an equation was developed to improve the SCS-CN method by combining the CN value with the tabulated CN_2 value and three introduced factors (slope gradient, soil moisture, and storm duration). Six models including the original SCS-CN method, Huang et al. [15], Huang et al. [34], and the proposed method with different optimized parameters (Methods 1–3) were used to test the reliability of the data from three runoff plots in the XDG watersheds of Loess Plateau. Subsequently, using the parameters calibrated and validated by dataset of the initial three runoff plots, the proposed method was then applied to runoff estimation of the remaining three runoff plots in CBG watershed. High NSE and low $RMSE$ values during the calibration, validation, and application period indicated the proposed method can estimate runoff accurately for six plots in two watersheds and had greater reliability than the standard SCS-CN, Huang et al. [15], and Huang et al. [34] methods. Moreover, Method 2 with an initial abstraction ratio of $\lambda = 0.001$ rather than the standard value of 0.2 and slope parameters obtained from Huang et al. [15] with data from slopes ranging from 14% to 140% seems to be the most promising SCS method for runoff estimation in the experimental plots of the Loess Plateau. Furthermore, the parameters of soil moisture (b_1 and b_2) are the most sensitive, followed by parameters in storm duration (c) and slope equations (a_1 and a_2), and the least parameter is the initial abstraction ratio λ on the basis of the proposed method sensitivity analysis.

Author Contributions: Conceptualization, N.W.; Data curation, N.W.; Methodology, W.S.; Writing—original draft, W.S. All authors have read and agreed to the published version of the manuscript.

Funding: This study was supported by the China Postdoctoral Science Foundation (2019M663917XB).

Acknowledgments: The authors thank the assistant editor and three anonymous referees for their useful advice and suggestions.

Conflicts of Interest: The authors declare no conflict of interest.

Data Availability Statement: The data that support the findings of this study are available on request from the corresponding author.

References

1. Shi, Z.H.; Chen, L.D.; Fang, N.F.; Qin, D.F.; Cai, C.F. Research on the SCS-CN initial abstraction ratio using rainfall-runoff event analysis in the Three Gorges Area, China. *Catena* **2009**, *77*, 1–7. [CrossRef]
2. SCS. *National Engineering Handbook, Section 4*; Soil Conservation Service USDA: Washington, DC, USA, 1972.
3. Shi, W.H.; Huang, M.B.; Barbour, S.L. Storm-based CSLE that incorporates the estimated runoff for soil loss prediction on the Chinese Loess Plateau. *Soil Tillage Res.* **2018**, *180*, 137–147. [CrossRef]
4. Shi, W.H.; Huang, M.B.; Wu, L.H. Prediction of storm-based nutrient loss incorporating the estimated runoff and soil loss at a slope scale on the Loess Plateau, China. *Land Degrad. Dev.* **2018**, *29*, 2899–2910. [CrossRef]
5. Kaffas, K.; Hrissanthou, V. Estimate of continuous sediment graphs in a basin, using a composite mathematical model. *Environ. Process.* **2015**, *2*, 361–378. [CrossRef]
6. Baginska, B.; Milne-Home, W.; Cornish, P.S. Modelling nutrient transport in Currency Creek, NSW with AnnAGNPS and PEST. *Environ. Model. Softw.* **2003**, *18*, 801–808. [CrossRef]
7. Knisel, W.G. CREAMS: A field-scale model for chemical, runoff and erosion from agricultural management systems. In *Conservation Research Report, Volume 26*; South East Area, US Department of Agriculture: Washington, DC, USA, 1980.
8. Sharpley, A.N.; Williams, J.R. *EPIC-Erosion/Productivity Impact Calculator: 1. Model Documentation*; U.S. Department of Agriculture Technical Bulletin No. 1768; U.S. Government Printing Office: Washington, DC, USA. 1990.
9. Arnold, J.G.; Moriasi, D.N.; Gassman, P.W.; Abbaspour, K.C.; White, M.J.; Srinivasan, R.; Santhi, C.; Harmel, R.D.; Van Griensven, A.; Van Liew, M.W. SWAT: Model use, calibration and validation. *Trans. ASABE* **2012**, *55*, 1491–1508. [CrossRef]
10. Petroselli, A.; Grimaldi, S. Design hydrograph estimation in small and fully ungauged basins: A preliminary assessment of the EBA4SUB framework. *J. Flood Risk Manag.* **2018**, *11*, S197–S210. [CrossRef]
11. Vojtek, M.; Petroselli, A.; Vojteková, J.; Asgharinia, S. Flood inundation mapping in small and ungauged basins: Sensitivity analysis using the EBA4SUB and HEC-RAS modeling approach. *Hydrol. Res.* **2019**, *50*, 1002–1019. [CrossRef]
12. Ponce, V.M.; Hawkins, R.H. Runoff curve number: Has it reached maturity? *J. Hydrol. Eng. ASCE* **1996**, *1*, 11–19. [CrossRef]
13. Bhuyan, S.J.; Mankin, K.R.; Koelliker, J.K. Watershed-scale AMC selection for hydrologic modeling. *Trans. ASABE* **2003**, *46*, 237–244. [CrossRef]
14. Babu, P.S.; Mishra, S.K. Improved SCS-CN–inspired model. *J. Hydrol. Eng.* **2012**, *17*, 1164–1172. [CrossRef]
15. Huang, M.; Gallichand, J.; Wang, Z.; Goulet, M. A modification to the Soil Conservation Service curve number method for steep slopes in the Loess Plateau of China. *Hydrol. Process.* **2006**, *20*, 579–589. [CrossRef]

16. Grimaldi, S.; Petroselli, A.; Romano, N. Green-Ampt Curve-Number mixed procedure as an empirical tool for rainfall-runoff modelling in small and ungauged basins. *Hydrol. Process.* **2013**, *27*, 1253–1264. [CrossRef]

17. Jain, M.K.; Mishra, S.K.; Babu, P.S.; Venugopal, K.; Singh, V.P. Enhanced runoff curve number model incorporating storm duration and a nonlinear Ia-S relation. *J. Hydrol. Eng.* **2006**, *11*, 631–635. [CrossRef]

18. Mishra, S.K.; Jain, M.K.; Babu, P.S.; Venugopal, K.; Kaliappan, S. Comparison of AMC-dependent CN-conversion formulae. *Water Resour. Manag.* **2008**, *22*, 1409–1420. [CrossRef]

19. Singh, P.K.; Mishra, S.K.; Berndtsson, R.; Jain, M.K.; Pandey, R.P. Development of a modified SMA based MSCS-CN model for runoff estimation. *Water Resour. Manag.* **2015**, *29*, 4111–4127. [CrossRef]

20. Hawkins, R.H. The importance of accurate curve numbers in the estimation of storm runoff. *Water Resour. Bull.* **1975**, *11*, 887–891. [CrossRef]

21. Hawkins, R.H. Asymptotic determination of runoff curve numbers from data. *J. Irrig. Drain. Eng. ASCE* **1993**, *119*, 334. [CrossRef]

22. Jiao, P.; Xu, D.; Wang, S.; Yu, Y.; Han, S. Improved SCS-CN method based on storage and depletion of antecedent daily precipitation. *Water Resour. Manag.* **2015**, *29*, 4753–4765. [CrossRef]

23. Michel, C.; Andréassian, V.; Perrin, C. Soil conservation service curve number method: How to mend a wrong soil moisture accounting procedure? *Water Resour. Res.* **2005**, *41*. [CrossRef]

24. Sahu, R.K.; Mishra, S.K.; Eldho, T.I. An improved AMC-coupled runoff curve number model. *Hydrol. Process.* **2010**, *24*, 2834–2839. [CrossRef]

25. Shi, W.H.; Huang, M.B.; Gongadze, K.; Wu, L.H. A modified SCS-CN method incorporating storm duration and antecedent soil moisture estimation for runoff prediction. *Water Resour. Manag.* **2017**, *31*, 1713–1727. [CrossRef]

26. Petroselli, A.; Grimaldi, S.; Piscopia, R.; Tauro, F. Design hydrograph estimation in small and ungauged basins: A comparative assessment of event based (EBA4SUB) and continuous (COSMO4SUB) modelling approaches. *Acta Sci. Pol. Form. Circumiectus* **2019**, *18*, 113–124.

27. Sahu, R.K.; Mishra, S.K.; Eldho, T.I. Improved storm duration and antecedent moisture condition coupled SCS-CN concept-based model. *J. Hydrol. Eng.* **2012**, *17*, 1173–1179. [CrossRef]

28. Hawkins, R.H. Runoff curve numbers with varying site moisture. *J. Irrig. Drain. Div.* **1978**, *104*, 389–398. [CrossRef]

29. Wood, E.F. An analysis of the effects of parameter uncertainty in deterministic hydrologic models. *Water Resour. Res.* **1976**, *12*, 925–932. [CrossRef]

30. Michele, C.D.; Salvadori, G. On the flood frequency distribution: Analytical formulation and the influence of antecedent soil moisture condition. *J. Hydrol.* **2002**, *262*, 245–258. [CrossRef]

31. Saxton, K.E. *User's Manual for SPAW: A Soil-Plant-Atmosphere Water Model*; USDA-ARS: Pullman, WA, USA, 1992.

32. Koelliker, J.K. *User's Manual of Potential Yield Model Revised (POTYLDR)*; Kansas State University, Department of Biological and Agricultural Engineering: Manhattan, KS, USA, 1994.

33. Jacobs, J.M.; Myers, D.A.; Whitfield, B.M. Improved rainfall/runoff estimates using remotely sensed soil moisture. *J. Am. Water. Resour. Assoc.* **2003**, *39*, 313–324. [CrossRef]

34. Huang, M.; Gallichand, J.; Dong, C.; Wang, Z.; Shao, M. Use of soil moisture data and curve number method for estimating runoff in the Loess Plateau of China. *Hydrol. Process.* **2007**, *21*, 1471–1481. [CrossRef]

35. Nash, J.E.; Sutcliffe, J.E. River flow forecasting through conceptual models: Part I-A discussion of principles. *J. Hydrol.* **1970**, *10*, 282–290. [CrossRef]

36. Risse, L.M.; Nearing, M.A.; Savabi, M.R. Determining the Green-Ampt effective hydraulic conductivity from rainfall-runoff data for the WEEP model. *Trans. ASAE* **1994**, *37*, 411–418. [CrossRef]

37. Liu, B.Y.; Nearing, M.A.; Risse, L.M. Slope gradient effects on soil loss for steep slopes. *Trans. ASAE* **1994**, *37*, 1835–1840. [CrossRef]

38. Kimura, R.; Liu, Y.; Takayama, N.; Zhang, X.; Kamichika, M.; Matsuoka, N. Heat and water balances of the bare soil surface and the potential distribution of vegetation in the Loess Plateau, China. *J. Arid Environ.* **2005**, *63*, 439–457. [CrossRef]

39. Marquardt, D.W. An algorithm for least-squares estimation of nonlinear parameters. *J. Soc. Ind. Appl. Math.* **1963**, *11*, 431–441. [CrossRef]

40. Ritter, A.; Muñoz-Carpena, R. Performance evaluation of hydrological models: Statistical significance for reducing subjectivity in goodness-of-fit assessments. *J. Hydrol.* **2013**, *480*, 33–45. [CrossRef]

41. Baltas, E.A.; Dervos, N.A.; Mimikou, M.A. Technical Note: Determination of the SCS initial abstraction ratio in an experimental watershed in Greece. *Hydrol. Earth Syst. Sci.* **2007**, *11*, 1825–1829. [CrossRef]

42. Walega, A.; Amatya, D.M.; Caldwell, P.; Marion, D.; Panda, S. Assessment of storm direct runoff and peak flow rates using improved SCS-CN models for selected forested watersheds in the Southeastern United States. *J. Hydrol. Reg. Stud.* **2020**, *27*, 100645. [CrossRef]

43. Hong, N. *Products and Servicing Solution Teaching Book for SPSS of Windows Statistical*; Tsinghua University Press, and Beijing Communication University Press: Beijing, China, 2003; pp. 300–311.

44. Wei, W.; Chen, L.L.; Fu, B.J.; Huang, Z.L.; Wu, D.P.; Gui, L.D. The effect of land uses and rainfall regimes on runoff and soil erosion in the semi-arid loess hilly area, China. *J. Hydrol.* **2007**, *335*, 247–258. [CrossRef]

45. Fang, H.Y.; Cai, Q.G.; Chen, H.; Li, Q.Y. Effect of rainfall regime and slope on runoff in a gullied loess region on the Loess Plateau in China. *Environ. Manag.* **2008**, *42*, 402–411. [CrossRef]

46. Mishra, S.K.; Pandey, R.P.; Jain, M.K.; Singh, V.P. A rain duration and modified AMC-dependent SCS-CN procedure for long duration rainfall-runoff events. *Water Resour. Manag.* **2008**, *22*, 861–876. [CrossRef]

47. Reaney, S.M.; Bracken, L.J.; Kirkby, M.J. Use of the connectivity of runoff model (CRUM) to investigate the influence of storm characteristics on runoff generation and connectivity in semi-arid areas. *Hydrol. Process.* **2010**, *21*, 894–906. [CrossRef]

48. Evett, S.R.; Dutt, G.R. Length and slope effects on runoff from sodium dispersed, compacted earth microcatchments. *Soil Sci. Soc. Am. J.* **1985**, *49*, 734–738. [CrossRef]

49. Philip, J.R. Hillslope infiltration: Planar slopes. *Water Resour. Res.* **1991**, *27*, 109–117. [CrossRef]

50. Huang, C.H. Empirical analysis of slope and runoff for sediment delivery from interrill areas. *Soil Sci. Soc. Am. J.* **1995**, *59*, 982–990. [CrossRef]

51. Fox, D.M.; Bryan, R.B.; Price, A.G. The influence of slope angle on final infiltration rate for interrill conditions. *Geoderma* **1997**, *80*, 181–194. [CrossRef]

52. Chaplot, V.A.M.; Bissonnais, Y.L. Runoff features for interrill erosion at different rainfall intensities, slope lengths, and gradients in an agricultural loessial hillslope. *Soil Sci. Soc. Am. J.* **2003**, *67*, 844–851. [CrossRef]

53. Koelliker, J.K. Hydrologic design criteria for dams with equal reliability. In *Irrigation Systems in the 21st Century*; James, L.G., Ed.; ASCE: New York, NY, USA, 1987; pp. 100–107.

54. Hjelmfelt, A.T.J.; Kramer, K.A.; Burwell, R.E. Curve numbers as random variables. In *Proceeding, International Symposium on Rainfall-Runoff Modelling*; Singh, V.P., Ed.; Water Resources Publication: Littleton, CO, USA, 1982; pp. 365–373.

Article

Variability of the Initial Abstraction Ratio in an Urban and an Agroforested Catchment

Adam Krajewski [1,*], Anna E. Sikorska-Senoner [2], Agnieszka Hejduk [3] and Leszek Hejduk [1]

[1] Department of Water Engineering and Applied Geology, Warsaw University of Life Sciences–SGGW, Nowoursynowska 159, 02-776 Warsaw, Poland; leszek_hejduk@sggw.pl

[2] Department of Geography, University of Zurich, Winterthurerstrasse 190, 8057 Zürich, Switzerland; anna.sikorska@geo.uzh.ch

[3] Water Center of Warsaw University of Life Sciences–SGGW, Ciszewskiego 6, 02-776 Warsaw, Poland; agnieszka_hejduk@sggw.pl

* Correspondence: adam_krajewski1@sggw.pl

Received: 28 November 2019; Accepted: 31 January 2020; Published: 4 February 2020

Abstract: The Curve Number method is one of the most commonly applied methods to describe the relationship between the direct runoff and storm rainfall depth. Due to its popularity and simplicity, it has been studied extensively. Less attention has been given to the dimensionless initial abstraction ratio, which is crucial for an accurate direct runoff estimation with the Curve Number. This ratio is most often assumed to be equal to 0.20, which was originally proposed by the method's developers. In this work, storm events recorded in the years 2009–2017 in two small Polish catchments of different land use types (urban and agroforested) were analyzed for variability in the initial abstraction ratio across events, seasons, and land use type. Our results showed that: (i) estimated initial abstraction ratios varied between storm events and seasons, and were most often lower than the original value of 0.20; (ii) for large events, the initial abstraction ratio in the catchment approaches a constant value after the rainfall depth exceeds a certain threshold value. Thus, when using the Soil Conservation Service-Curve Number (SCS-CN) method, the initial abstraction ratio should be locally verified, and the conditions for the application of the suggested value of 0.20 should be established.

Keywords: Curve Number; initial abstraction ratio; direct runoff; SCS-CN method; small catchment

---

1. Introduction

Estimation of the direct runoff as a catchment response to a rainfall or snowmelt event is important for many practical applications: The direct runoff is the key variable used for designing hydraulic structures [1], estimating soil erosion and sediment yield [2], or analyzing long-term changes in water resources availability [3]. The amount of water outflowing after a storm event depends on, among others, the catchment topography, land cover, soil characteristics, and climate conditions. Hence, for a specific site, the amount of direct runoff may vary significantly over successive years but also from one storm event to another within the same year [4]. This relationship between the direct runoff and the storm rainfall depth may be assumed constant given the same rainfall depth and unchanged catchment conditions.

One of the most commonly applied methods to describe this relationship between the direct runoff and storm rainfall depth is the Soil Conservation Service-Curve Number (SCS-CN), developed by the United States Department of Agriculture (USDA) [5]. Despite its simplicity, this method accounts for most of the catchment runoff characteristics responsible for producing the direct runoff such as soil type; land use and treatment; and surface and antecedent moisture conditions. Moreover, its parameter (i.e., curve number, CN) can be derived directly from physical properties of the catchment or from recorded rainfall-runoff events [6]. Therefore, the SCS-CN method remains very attractive for runoff estimation in poorly gauged regions, in applied hydrology for engineering designs, or in post analysis of recorded data [7,8], where the simplicity and the universality of the method play a deciding role. As this method estimates only the direct runoff without considering the base flow component, it is most often applied to event-based models [9]. According to SCS-CN, the relationship between the rainfall depth, direct runoff, and catchment retention can be described as follows:

$$\frac{H}{P - Ia} = \frac{F}{S} \tag{1}$$

where:

    F: actual catchment retention (mm),
    S: potential maximum catchment retention (mm),
    H: direct runoff, effective rainfall (mm),
    P: rainfall depth (mm),
    Ia: initial abstraction (mm).

Note that, $F \leq S$ and $H \leq (P - Ia)$. The total retention for a storm consists of both Ia and F, so the conservation of the mass Equation can be expressed as:

$$F = P - Ia - H \tag{2}$$

Substituting Equation (2) for F in Equation (1) and solving it for the direct runoff results in:

$$H = \frac{(P - Ia)^2}{P - Ia + S} \tag{3}$$

The initial abstraction (Ia) consists of evaporation, interception, infiltration, and surface depression storage. If rainfall depth is lower or equal to the initial abstraction, the direct runoff is assumed to be zero (i.e., all rainfall water is used to fill the catchment retention). According to the National Engineering Handbook [5], an empirical relationship between Ia and S has been established as:

$$Ia = \lambda \times S \tag{4}$$

where λ is the initial abstraction ratio (-).

The potential maximum catchment retention (S) is next calculated from the following formula:

$$S = 25.4\left(\frac{1000}{CN} - 10\right) \tag{5}$$

where CN is curve number, i.e., the only parameter (-) of the SCS-CN method that depends on soil characteristics, land use, and catchment initial conditions prior to the storm event.

Due to its popularity, the SCS-CN method has been the object of many studies. These studies focused on application of the method, e.g., in rainfall-runoff modelling [10], analyzing the impact of land use changes such as urbanization on runoff values [11], predicting sediment yield [12,13], as well as on the estimation and verification of the CN parameter [14–17]. Yet, less attention has been given to the dimensionless initial abstraction ratio (λ), and most often this ratio was assumed equal to 0.20, as it was originally proposed by method's developers [5]. Although Woodward et al. [18] have suggested based on intensive data of 327 US catchments that a value of 0.05 may be more appropriate than 0.2, most of the applied studies continued to use the originally proposed value of 0.2. Recently, this dimensionless initial abstraction ratio (λ) has gained much more attention and several researchers have revealed that (a) this ratio may vary widely for specific sites [19,20], and (b) it may be significantly lower than 0.20 [18,21,22]. A change in the initial abstraction ratio from a fixed value of 0.20 to a lower value will strongly affect the estimated depth of the direct runoff and consequently the direct runoff distribution over time. Durán-Barroso et al. [23] have even stated this relationship between the initial abstraction and the maximum potential catchment retention as being one of the major sources of uncertainty in the SCS-CN method. Hence, some researchers have tried to estimate the value of this ratio from recorded storm data, and they concluded that using a varying value of the initial abstraction ratio rather than a fixed value of 0.2 or even 0.05 significantly improves runoff estimates [9,20,23]. Nevertheless, the number of studies devoted to the initial abstraction ratio is still very limited. As this ratio was also declared as being regional specific [18], more research is needed for catchments in different climatic regions to investigate properties of this ratio.

Another important aspect of the initial abstraction ratio is linked with the basic assumption of the SCS-CN method that both this ratio and the *CN* parameter are assumed constant given that catchment conditions do not change. Yet, any change in the catchment conditions will affect the relationship between the storm rainfall depth and the direct runoff amount, and consequently the value of the *CN* parameter and possibly the initial abstraction ratio. In this respect, direct human activities such as land use changes and human induced climate changes are considered as two major factors that affect runoff processes the most. Many studies have tried to quantify their role in the hydrological cycle [24–26]. For instance, Chen et al. [27] revealed that for a large basin (area = 18,827 km^2) in northwestern China, climate variability was the major factor affecting the runoff amount, accounting for 90.5% of the increase in the runoff amount, while other human activities accounted only for 9.5% of the runoff increase. In contrast to that, Lv et al. [28] found for the same region that in the case of a small catchment (area = 72.8 km^2), climate change accounted only for about 33% of the total decrease in mean annual runoff, while direct human activity was responsible for about 67% of the decrease in mean annual runoff. Tan and Gan [29], based on investigations in 96 Canadian catchments, showed that climate change generally caused an increase in the mean annual runoff, while direct human impacts usually lead to decreasing water resources. These studies indicate how complex interactions between climatic, anthropogenic, and hydrological variables are, and how difficult it is to evaluate their individual impacts on the hydrological cycle in the catchment [30]. Although these climatic and direct anthropogenic factors can be accounted for in the SCS-CN method by simply adjusting the curve number (*CN*), it remains unclear whether and how exactly these factors may affect the value of the initial abstraction ratio in the SCS-CN method.

In this work, recorded rainfall-runoff data collected in the years 2009–2017 in two small Polish catchments of different land use types (urban and agroforested) were analyzed in order to a) estimate and verify the value of the initial abstraction ratio in these two catchments for different storm events, b) analyze the variation of the initial abstraction ratio with the rainfall depth, and c) explore (indirectly) the effect of the land use type and season on the initial abstraction ratio value in this climatic region. The main focus of this study is to investigate such local estimates of the initial abstraction ratio based on limited recorded storm events for two small catchments of different land use types. This study is a follow-up of two previous studies conducted for these two catchments [14,31] that investigated the applicability of the SCS-CN method and estimated the *CN* parameter values based on conventional methods. Therefore, the estimation of the *CN* and its verification are not the main subjects of this study.

2. Materials and Methods

2.1. Characteristics of the Study Catchments

Data analyzed in this work were collected in two small catchments of different land use types: urban at Służew Creek and agroforested at Zagożdżonka River (Figure 1). In both of them, meteorological and hydrological variables are automatically monitored by the Department of Water Engineering and Applied Geology (DoWEaAG) of Warsaw University of Life Sciences-SGGW (WULS-SGGW). Table 1 presents basic topographic and climatic characteristics of both study sites. Both of these catchments are under human pressure. In the urban catchment of Służew Creek, for the last 40 years an increase of impervious areas has been observed, which, among others, leads to a higher flood risk [32]. In the agroforested catchment of Zagożdżonka River, decreasing trends in low and medium discharges have been observed [3]. Research on the rainfall-runoff process under the above mentioned conditions is particularly valuable as it helps to manage water resources in a changing environment.

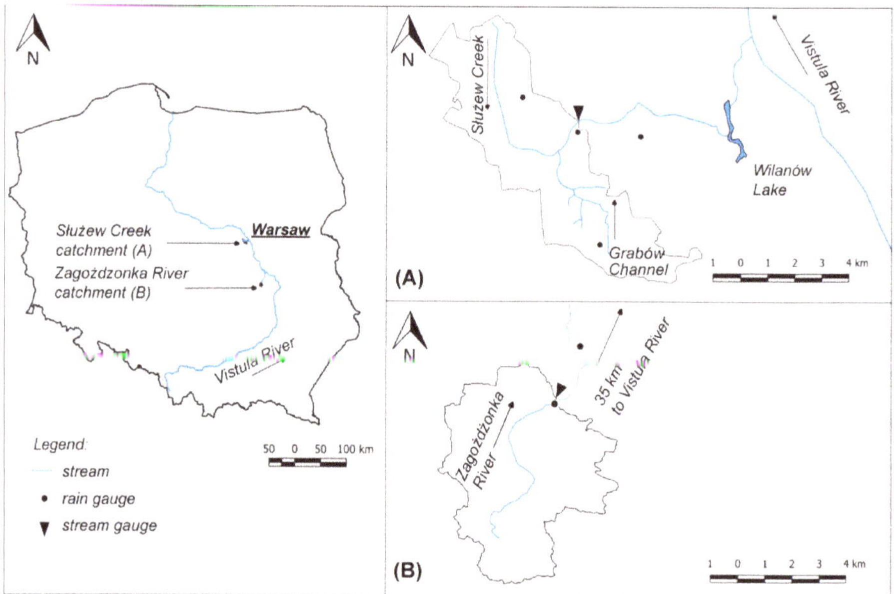

Figure 1. Locality map of Służew Creek (**A**) and Zagożdżonka River (**B**) catchments up to the Wyścigi and Czarna gauging stations, respectively.

The first catchment, Służew Creek, is located in the southwestern part of Warsaw in Poland. Its total area equals 54.8 km^2, whereas the subcatchment up to Wyścigi gauging station (analyzed in this work) has an area of 28.7 km^2. This catchment has heterogeneous land use (see Table 2 for details) with more than 60% of its area covered by built-up areas and with the impervious factor of the total catchment reaching up to 25%. Fine-grained soils dominate the watershed area. Previous investigations conducted in the Służew Creek catchment were related to urban hydrology with a focus on flood flow estimation [33,34] and operation of small detention ponds for flood and sediment load reduction [35,36] or sediment transport [37,38]. The climate in Poland is continental and in the case of the analyzed study site it may be described as moderate. Long-term measurements indicate that the average annual precipitation is 541.9 mm [39] and the average annual temperature is 8.6 °C [40]. The average annual discharge at Wyścigi gauging station in the analyzed period (2009–2017) was 0.182 m^3 s^{-1} [41].

The second analyzed catchment, Zagożdżonka River, is a left tributary of the Vistula River. Its catchment is located in central Poland, about 100 km south of Warsaw (Figure 1). Hydrological research in the upper part of the Zagożdżonka River has been conducted by WULS-SGGW since the 1960s. It was initiated to determine the water needs of the chemical plant. Over time, the scope of investigation has expanded [2,4,10,31], and currently there are two stream gauging stations, at Płachty Stare (since 1962) and Czarna (since 1991), at which measurements of basic meteorological and hydrological parameters are carried out continuously. In contrast to the urban catchment mentioned above, the Zagożdżonka River catchment can be defined as agricultural and forested. The catchment area at Czarna (considered in this study) is equal to 23.4 km^2. Forests (45%) and arable lands (43%) cover a significant part of the catchment area (see Table 2 for details). Sandy soils are the major soil type in this catchment, while in local depressions and river flood plains, peaty soils may be found.

Table 1. Topographic and climatic characteristic of investigated catchments.

Category	Unit	Służew Creek (Years) [Source]	Zagożdżonka River (Years) [Source]
Catchment type	-	lowland, urbanized	lowland, agroforested
Area up to gauging station	km^2	28.7 [41]	23.4 [41]
Average channel slope	‰	1.1 [41]	3.0 [41]
Impervious surfaces	%	25.3 [42]	0.56 [42]
Average precipitation	mm	541.9 (1960–2009) [39]	612.0 (1963–2015) [3]
Average temperature	°C	8.6 (1970–2009) [40]	8.1 (1951–2015) [3]
Average discharge per sq. km	dm^3·s^{-1}·km^{-2}	6.34 (2010–2017) [41]	3.89 (2010–2017) [41]

Table 2. Land cover and soil types in the Służew Creek catchment up to Wyścigi gauging station and in the Zagożdżonka River catchment up to Czarna station.

Category	Land Cover (%) and Soil Structure (%) in	
	Służew Creek	Zagożdżonka River
Land cover type [42]		
Residential areas	28.9	3.6
Airport	19.3	-
Arable lands	12.9	43.4
Forests	12.5	45.1
Industrial areas	9.4	-
Recreation areas	7.9	-
Roads and associated lands	7.1	-
Meadows	2.0	7.8
Soil type [43]		
Loamy sand	50.5	-
Light loamy sand	40.2	-
Organic soil	9.3	-
Silt loam	-	67.8
Silt	-	27.3
Sand	-	4.9

The relief in both catchments is typical for lowlands in Poland with an insignificant elevation variation. Thus, the catchment elevation does not play any relevant role in the runoff generation in these two catchments.

2.2. Recorded Data

Hydrological field investigations were carried out in both catchments according to the standard methodology recommended by the Polish Institute of Meteorology and Water Management, National Research Institute [44,45]. Rainfall depths were measured with the use of tipping-bucket rain gauges at four locations in the Służew Creek catchment and two in the Zagożdżonka River catchment. Recorded rainfall depths were summed up in 10 min intervals and the average rainfall over the catchment was calculated according to the Thiessen polygons method. River gauging stations at catchments outlets were equipped with staff gages and automatic, pressure water stage loggers. At these river gauging stations, water stages data were collected every 10 min. In order to estimate and verify the automated measurements, readings of the gage staff were regularly performed at least once a week. Recorded water levels were next converted to discharge values with the help of the rating curve method [46]. The rating curves were established at both sites with numerous hydrometric measurements, i.e., cross section area and flow velocity, conducted at low, medium, and high stages. These curves were verified each year to reduce the uncertainty in the rating curve method [3].

2.3. Selection of Storm Rainfall-Runoff Events

From all rainfall events recorded in the years 2009–2017, we selected only those events that generated a direct runoff. All of these events were caused by rainfall events in the spring-summer seasons. This criterion led to a selection of 53 storm events in the Służew Creek, and 42 storm events in the Zagożdżonka River, respectively, with an average direct runoff of about 3 mm (2.9 mm in Zagożdżonka River). The average rainfall depth of selected events was equal to 22.8 mm in the Służew Creek and to 25.8 mm in the Zagożdżonka River.

Antecedent moisture conditions (AMC) were estimated for all 95 events (see Table 3). Within these events, 85 were classified as AMC I, seven as AMC II, and three as AMC III. Therefore, it can be stated that the majority of recorded events occurred during dry conditions. This finding is surprising because the analyzed period 2009–2017 was not extremely dry or less rich in water compared to previous years [47]. Thus, it is likely that the thresholds used to determine moisture conditions may not correspond to the climate of central Poland and should be adapted for this region in a future study.

Table 3. Antecedent moisture condition (AMC) for analyzed events according to the Soil Conservation Service-Curve Number method.

Antecedent Moisture Condition	Total Number of Events with 5 Days of Rainfall for	
	Dormant Period	**Vegetation Period**
AMC I	(12 events) < 13 mm	(73 events) < 35 mm
AMC II	13 mm ≤ (3 events) ≤ 28 mm	35 mm ≤ (4 events) ≤ 53 mm
AMC III	(1 event) >28 mm	(2 events) > 53 mm

2.4. Estimation of the Initial Abstraction Ratio from Recorded Data

The initial abstractions were estimated based on recorded storm events (see Figures 2 and 3) as the sum of the rainfall depth after which a direct runoff begins (seen as an increase in water stage and river discharge). The volume of the direct runoff was calculated based on the measured hydrograph, e.g., by indicating the event start point (increase in river discharge after the rainfall begins), the event end point (change in the slope of the hydrograph falling limb), and by removing the base flow from the total runoff. The base flow was estimated following the straight-line separation method that is described in detail by Wanielista et al. [48]. The computed volume of the direct runoff was next divided by the catchment area, giving the runoff depth for each storm event. As the rainfall depth, the direct runoff, and the initial abstraction are known (measured or estimated), Equation (3) may be solved for the maximum potential catchment retention: S. Finally, the initial abstraction ratio was computed from Equation (4) for each recorded storm event. This technique of Ia estimation has been used in previous studies such as Shi et al. [8] and Ling et al. [49].

The estimated values for the initial abstraction ratio were compared with the original value of 0.2 that was suggested in the original version of the SCS-CN method [5].

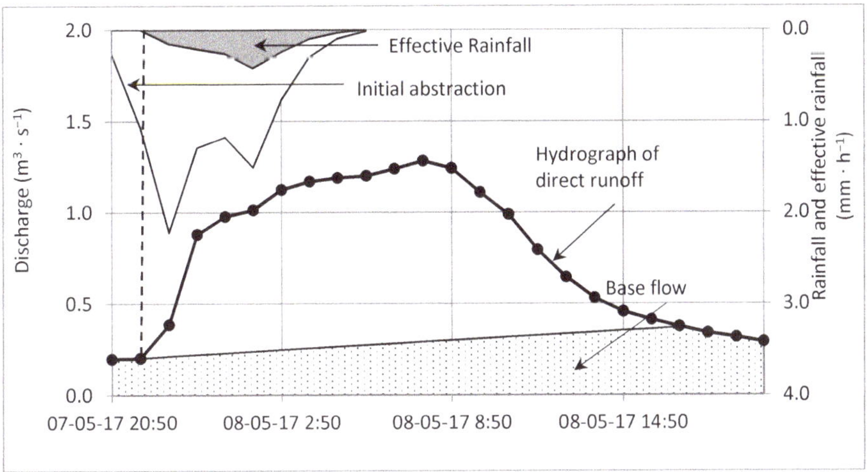

Figure 2. Example of the rainfall-runoff event, 7 May 2017, recorded in the Służew Creek catchment.

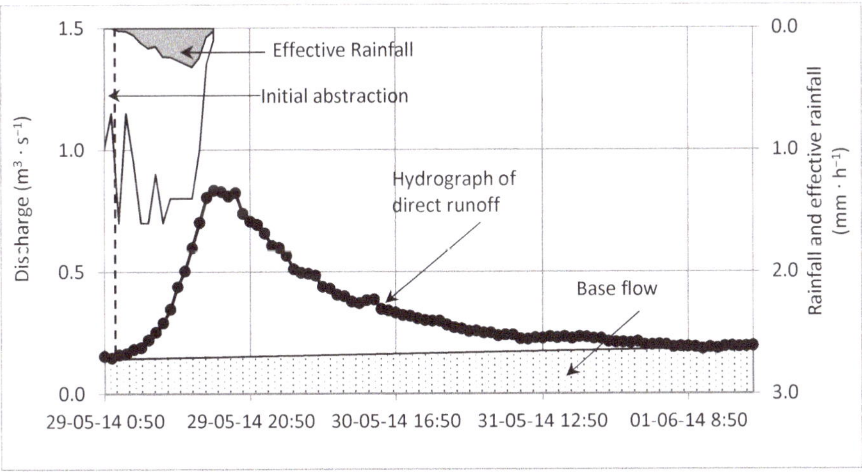

Figure 3. Example of the rainfall-runoff event, 29 May 2014, recorded in the Zagożdżonka River catchment.

2.5. Variability of the Initial Abstraction Ratio

In order to investigate seasonal variation in the initial abstraction ratio due to the state of the vegetation development and other seasonal factors, the recorded storm events were next grouped into three periods: (I) beginning of the vegetation season (April–May), (II) full vegetation state (June–August), and (III) end of the vegetation season and plants resting period (September–March). The analysis of the initial abstraction ratio was performed for all events in each group separately and the estimated values for each group were compared.

Finally, we investigated whether a relationship between the initial abstraction ratio and the rainfall depth can be established for our recorded data. The relationship between the *CN* parameter and the rainfall depth has been explored [14,31,50]. Among others, Soulis and Valiantzas [51,52] studied the effect of soils and land cover variability on the hydrologic response and proposed a methodology to determine the spatial distribution of the SCS-CN parameter values from rainfall-runoff data. However, similar analyses of the initial abstraction ratio have rarely been performed [8,20] and did not reveal any relationships between these two variables. In our case, we focused only on summer events (i.e., from May to September), to reduce the possible effect of the seasonal variability in the ratio in the studied catchments and also because most of the recorded events occurred during this period.

The relationship between the initial abstraction ratio and the rainfall depth has been approximated with an asymptotic formula:

$$\lambda(P) = \lambda_\infty + (1 - \lambda_\infty)exp\left(-\frac{P}{b}\right)$$

(6)

where:

λ_∞: initial abstraction ratio (-) for rainfall tending to infinity,

P: rainfall depth (mm),

b: constant parameter (-).

Parameters in Equation (5) were next identified using the Table Curve 2D software [53].

3. Results and Discussion

3.1. Initial Abstraction Ratio Variation

Characteristics of the selected storm events for both catchments are summarized in Table 4. In the Służew Creek catchment, recorded rainfall depths ranged from 4.3 mm up to almost 76 mm, with the average runoff coefficient (ratio of rainfall to runoff) equal to 0.125. The estimated initial abstraction ratio (λ) for all recorded events varied from 0.002 to 0.188 and thus was lower than the original value of 0.20. The λ average value over all 53 analyzed storm events was equal to 0.026.

Table 4. Characteristics of rainfall-runoff events recorded in the catchments of Służew Creek and Zagożdżonka River.

Category	Abbr.	Unit	Służew Creek		Zagożdżonka River	
			Range	Average	Range	Average
No. of events	n	-	-	53	-	42
Rainfall depth	P	mm	4.3–75.8	22.8	5.8–145.1	25.8
Direct runoff	H	mm	0.2–16.3	3.0	0.2–21.8	2.9
Runoff coefficient	$c = H/P$	-	0.053–0.295	0.125	0.018–0.577	0.106
Initial abstraction	Ia	mm	0.1–4.8	2.2	0.4–12.5	3.7
Maximum potential catchment retention	S	mm	23.4–683.3	151.3	9.1–1342	280.4
Initial abstraction ratio	$\lambda = Ia/S$	-	0.002–0.188	0.026	0.001–0.512	0.047

In the case of the Zagożdżonka River catchment, rainfall depths varied from 5.8 mm to 145.1 mm, with the average runoff coefficient estimated as 0.106. The average initial abstraction ratio in the Zagożdżonka River catchment over all 42 storm events was equal to 0.047, which was almost twice as high as in the Służew Creek catchment but still lower than 0.20. For individual events, λ varied from 0.001 to 0.512. Among all analyzed storm events, only for two events was λ higher than 0.2, while for all other 40 events it was below this original value.

The values of λ for individual events are presented in Figure 4. In the case of the Zagożdżonka River, the data are more dispersed, meaning that the variation in λ for different storm events was higher than for the Służew Creek catchment. However, whereas the mean values varied for both catchments, the median values are similar, i.e., 0.017 and 0.018, respectively, for Służew Creek and Zagożdżonka River.

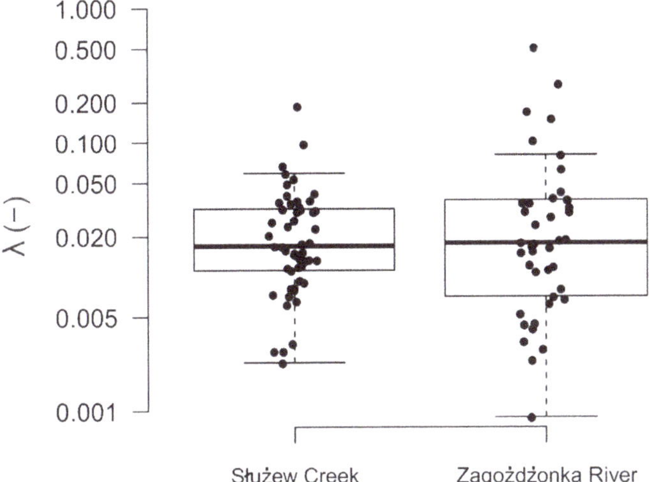

Figure 4. Distribution of initial abstraction ratios (λ) for two analyzed catchments; note the logarithmic scale.

These findings demonstrate that for both catchments the initial abstraction ratio λ varyied for different storm events, and for most storm events it was lower than the original value of 0.2 suggested by the authors of [5] and often even lower than 0.05, which was proposed as a modification of the originally developed method [18]. Similar results were reported by other researchers [9,20,23], who also used a varying value for λ in their studies. These findings suggest that using a varying value of λ may indeed be more appropriate than a constant fixed value for all events.

3.2. Initial Abstraction Ratio Dependency on the Land Use Type

The Zagożdżonka River catchment with 96% of undeveloped areas (forests, meadows, arable lands) has a much higher retention capacity than does the Służew Creek catchment, where undeveloped areas cover only 35%. Thus, as the elevations of these catchments do not play any role in the runoff generation, we can conclude that the value of λ depends on the land use type in the catchment and it decreases with an increasing ratio of the catchment imperviousness. Thus, it is logical that the value of λ is smaller than 0.2 in the urbanized catchment of Służew Creek. It is surprising, however, that even in the Zagożdżonka catchment that has a high proportion of forests and meadows (53%) the ratio λ did not reach the original value of 0.20.

3.3. Initial Abstraction Ratio Dependence on the Vegetation State

As it appears from our results in Section 3.2, the initial abstraction ratio can be related to the dominant land use in the catchment. Hence, its value may also change seasonally due to the state of the vegetation development. Figure 5 presents the variability of initial abstraction ratio within the three investigated seasons in both studied catchments, i.e., (I) beginning of the vegetation season (April–May); (II) full vegetation state (June–August); and (III) end of the vegetation season and plants resting period (September–March).

In the case of the Zagożdżonka River catchment a noticeable increase in the value of λ may be noticed for the vegetation period (May–September). During the full vegetation period, rainfall depths were intercepted by plants and initial abstractions were the greatest. For the Służew Creek catchment, no relevant change in the ratio was observed and the ratio value seemed to be relatively constant over time. The small percentage of vegetation in this catchment, which is dominated by the urban area instead, probably limits the role of the interception in the runoff formation. Hence, we can conclude that the seasonal variability in the initial abstraction ratio can be observed only in a catchment where vegetation plays a role. Depending on the vegetation type and its percentage in the catchment, this factor may need to be accounted for when estimating the initial abstraction ratio in agricultural or forested catchments. No variation in the ratio should be expected in urbanized or highly impervious catchments and an average value for the ratio over all seasons could be used.

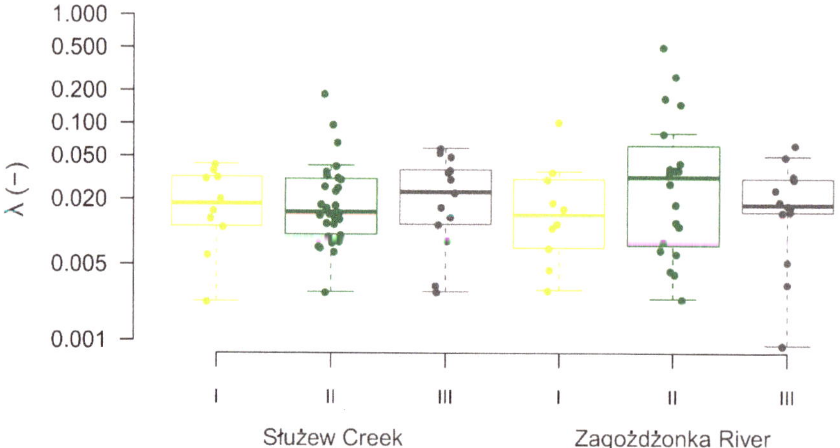

Figure 5. Changes in initial abstraction ratios within vegetation seasons for two analyzed catchments; note the logarithmic scale.

3.4. Initial Abstraction Ratio Variation with Rainfall Depth

In the case of the Służew Creek catchment, 43 storm events occurred during the summer period (May–September) and the estimated values of the initial abstraction ratio for these events have been plotted versus rainfall depths in Figure 6. Although the variability of λ with rainfall depth was high, for rainfall depths higher than 30 mm, the initial abstraction ratio seemed to be relatively constant and varied about the value of 0.0089.

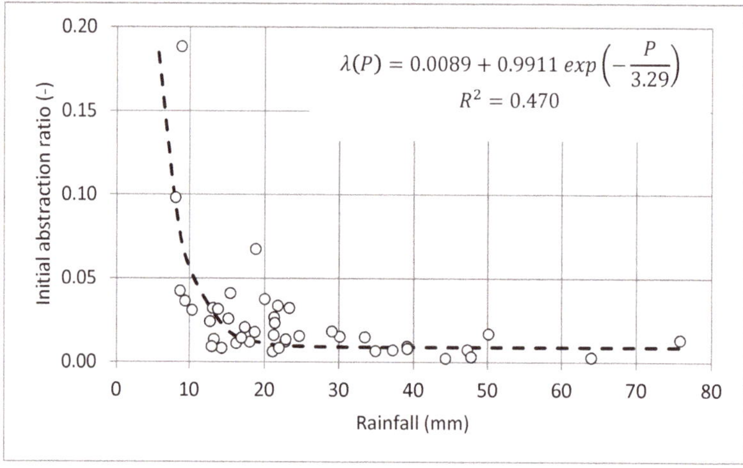

Figure 6. Initial abstraction ratio vs. rainfall depth for summer period, Służew Creek catchment.

The same procedure was repeated for the Zagożdżonka River, resulting in 32 summer events and the results are plotted in Figure 7. In this catchment, a greater variation of λ (especially for low rainfalls) was observed than for that in the Służew Creek catchment (compare with Figure 6). Although the relationship between the initial abstraction ratio and the rainfall depth was in this catchment less pronounced, a similar tendency to the Służew Creek catchment could also be observed here, i.e., a constant value of the initial abstraction ratio approaching 0.0109 for rainfalls with the rainfall depth higher than 20 mm may be noticed. For the case of the Służew Creek catchment, the coefficient of determination (R^2) equaled 0.470 and for the Zagożdżonka River catchment it was 0.236.

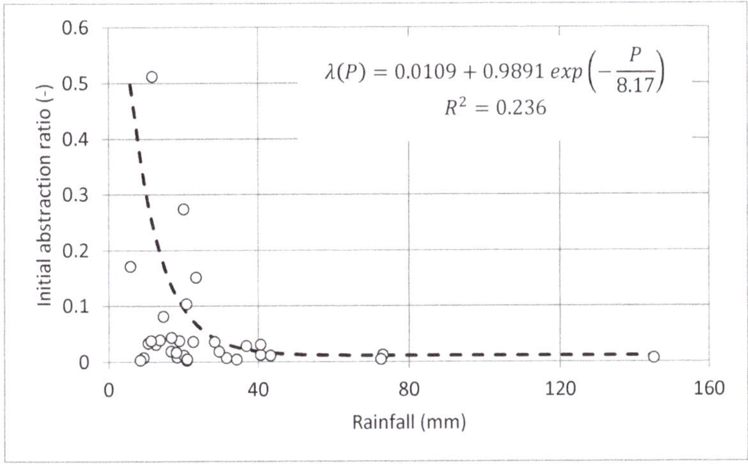

$$\lambda(P) = 0.0109 + 0.9891 \; exp\left(-\frac{P}{8.17}\right)$$
$$R^2 = 0.236$$

Figure 7. Initial abstraction ratio vs. rainfall depth for summer period, Zagożdżonka River catchment.

Estimations of similar relationships for the other two seasons were not performed due to too few events per season. The relationship between the initial abstraction and the rainfall depth, and the initial abstraction and the maximum potential retention, could not be established for recorded events (no relationship was observed).

A similar phenomenon with the equilibrium in the initial abstraction ratio was observed by Santikari and Murdoch [9] who also found that the initial abstraction amount approaches some constant value after a certain rainfall depth is exceeded. They, however, did not explore the relationship between the initial abstraction ratio and the rainfall depth itself. This phenomenon observed in our data can be explained by the fact that usually after heavy rainfalls, the entire catchment area takes part in the runoff formation. Therefore, the initial abstraction does not increase further and, consequently, the initial abstraction ratio approaches a constant value. In contrast to that, during small rainfalls, only some and often various parts of the catchment contribute to the runoff formation, causing a variation in the initial abstraction ratio. Although usually the catchment area located closest to the channel contributes most to the generated runoff during small rainfall events, the exact catchment contribution depends on the movement of the storm cloud and the detection of the rainfall cloud by local point measurements [54]. Note that the SCS method was originally designed for a direct runoff estimation of large (annual) events, during which the spatial patterns of the rainfall do not play any significant role in the runoff generation [5]. In reference to the above, Hawkins et al. [55] suggested selecting representative events in the watershed, i.e., with rainfall depths higher than 1 inch (25.4 mm), for analyses of the *CN* parameter. In our case, we did not limit the selection of events to rainfalls exceeding this threshold value, as this would reduce the number of analyzed events to 13 in Służew Creek and 12 in Zagożdżonka River. Application of the SCS-CN method to rainfall events smaller than annual events is a common practice for runoff estimation in ungauged catchments, where other more data intensive models cannot be applied [8,15,33]. Such applications to small(er) events may, however, result in a lower model performance than that expected by the method, and the reasons for that should always be clarified.

3.5. Initial Abstraction Ratio May Change Locally

As our results from recorded storm events in two catchments of different dominant land use type showed, the initial abstraction ratio varies from storm event to storm event, and it is related to the rainfall depth. This ratio also varies seasonally and is linked to the land use type as well as the state of the vegetation. It should also be noted that the method of *Ia* estimation applied for this study is devoted to small catchments, for which the effect of water travel time within the catchment may be neglected. The presented approach does not suit large catchments, where the effect of runoff routing and the effect of the spatial variability of rainfall on the runoff production may play a significant role. Thus, for such large size catchments, the validity of the obtained results may be limited. Hence, more research is needed on the initial abstraction ratio in different climatic regions and also for catchments with different land use types, to verify results found in our study and to compare local estimates with original values (0.2 and 0.05). We suggest that if local information is available, the initial abstraction ratio for the SCS-CN method should be estimated from locally collected data before applying the method to flood estimation or designing flood structures. Of course, estimating the value of λ for each event or catchment independently can be problematic in practice especially for ungauged catchments without recorded storm events. However, it should still be possible and reasonable to use an initial abstraction ratio value that is specific for the region or catchment type, if such values are available. In the case of no regional or catchment-type values being available, standard values as originally proposed in the SCS-CN method should be used instead. Such local verification is especially required if the SCS-CN method is applied outside of its original scope, i.e., to storm events smaller than annual floods. Using the originally proposed value of 0.2 for such small storm events may be linked with underestimation of the runoff amount and thus flood risk in catchments, where the estimated value of the initial abstraction ratio is much lower than its original value. If only large annual floods are of interest, we suggest using the value proposed originally, i.e., 0.2, as our results demonstrated the initial abstraction ratio may indeed produce values close to this original value under some conditions (e.g., large floods) and these conditions should be first verified locally to define the application boundaries for the original value of 0.2. For instance, one should define the return period of floods for which the original value of 0.2 or 0.05 could still be applied, and the threshold value of the return period above which local estimates are preferable.

4. Conclusions

In this work, rainfall-runoff storm events recorded in the period of 2009-2017 in two small Polish catchments of different land use types (urbanized and agroforested) were analyzed to estimate the initial abstraction ratio of the SCS-CN method and to explore its variability with rainfall depth, land use type, and season. Based on our results, the following conclusions have been made:

- Estimated values of the initial abstraction ratio varied between storm events and were most often lower than the original value of 0.20 originally suggested. For the urbanized catchment, this ratio ranged from 0.002 to 0.188, with the average value of 0.025, while for the agroforested catchment, the average ratio equaled 0.047, ranging from 0.001 to 0.512;
- The initial abstraction ratio varies over seasons, especially in the case of non-urbanized catchments with a high percentage of area covered with vegetation. Its higher value is expected during the vegetation period due to the occurrence of plants' interception. In urbanized catchments, a variation of the ratio over seasons is expected to be small;
- For large rainfall events, the initial abstraction ratio approaches a constant value after the rainfall depth exceeds a certain threshold value. This threshold is catchment dependent and in the case

of the urbanized catchment it was $p > 30$ mm ($\lambda \rightarrow 0.0089$) and in the case of the agroforested catchment it was $p > 20$ mm ($\lambda \rightarrow 0.0109$);

- When using the SCS-CN method, special care should be taken to determine the initial abstraction ratio, as the suggested value of 0.20 may be too high. This may result in underestimation of the runoff depth that may have serious consequences in terms of the underestimation of flood risk.

Author Contributions: A.K. and A.S. designed the study. A.K. performed the analysis for Służew Creek and wrote the first draft of the manuscript. A.H. and L.H. performed the analysis for Zagożdżonka River. A.S, A.H., and L.H. All authors have read and agreed to the published version of the manuscript.

Funding: This research was partly supported by the following projects: the grant from Island, Lichtenstein, and Norway through the EEA Financial Mechanism and the Norwegian Financial Mechanism, the research projects N N305 396238 and 2015/19/N/ST10/02665 founded by the National Science Center, Poland (NCN), and the research project KORANET EURRO-KPS founded by the National Center for Research and Development (NCBiR). The support provided by the institutions is gratefully acknowledged. The authors wish to thank the associate editor and two anonymous referees for providing useful comments that helped us to improve this paper.

Conflicts of Interest: The authors declare no conflict of interest.

References

1. Rabori, A.M.; Ghazavi, R. Urban Flood Estimation and Evaluation of the Performance of an Urban Drainage System in a Semi-Arid Urban Area Using SWMM. *Water Environ. Res.* **2018**, *90*, 2075–2082. [CrossRef] [PubMed]
2. Banasik, K.; Gorski, D.; Popek, Z.; Hejduk, L. Estimating the annual sediment yield of a small agricultural catchment in central Poland. In *Erosion and Sediment Yields in the Changing Environment*; Collins, A.V.G., Horowitz, A.X.L., Stone, M., Walling, D., Eds.; IAHS Publ.: Wallingford, UK, 2012; Volume 256, pp. 267–275.
3. Krajewski, A.; Sikorska-Senoner, A.E.; Ranzi, R.; Banasik, K. Long-Term Changes of Hydrological Variables in a Small Lowland Watershed in Central Poland. *Water* **2019**, *11*, 564. [CrossRef]
4. Banasik, K.; Hejduk, L. Long-term changes in runoff from a small agricultural catchment. *Soil Water Res.* **2012**, *7*, 64–72. [CrossRef]
5. National Engineering Handbook (NEH). *National Engineering Handbook, Part 630 Hydrology*; USDA: Washington, DC, USA, 2004.
6. Sikorska, A.E.; Scheidegger, A.; Banasik, K.; Rieckermann, J. Bayesian uncertainty assessment of flood predictions in ungauged urban basins for conceptual rainfall-runoff models. *Hydrol. Earth Syst. Sci.* **2012**, *16*, 1221–1236. [CrossRef]
7. Hawkins, R.H. Asymptotic Determination of Runoff Curve Numbers from Data. *J. Irrig. Drain. Eng.* **1993**, *119*, 334–345. [CrossRef]
8. Shi, Z.-H.; Chen, L.-D.; Fang, N.-F.; Qin, D.-F.; Cai, C.-F. Research on the SCS-CN initial abstraction ratio using rainfall-runoff event analysis in the Three Gorges Area, China. *Catena* **2009**, *77*, 1–7. [CrossRef]
9. Santikari, V.P.; Murdoch, L.C. Including effects of watershed heterogeneity in the curve number method using variable initial abstraction. *Hydrol. Earth Syst. Sci.* **2018**, *22*, 4725–4743. [CrossRef]
10. Banasik, K.; Hejduk, L.; Woodward, D.; Banasik, J. Flood peak discharge vs various CN and rain duration in a small catchment. *Rocz. Ochr. Srodowiska* **2016**, *18*, 201–212.
11. Li, C.; Liu, M.; Hu, Y.; Shi, T.; Zong, M.; Walter, M.T. Assessing the Impact of Urbanization on Direct Runoff Using Improved Composite CN Method in a Large Urban Area. *Int. J. Environ. Res. Public Health* **2018**, *15*, 775. [CrossRef]
12. Singh, P.; Bhunya, P.; Mishra, S.; Chaube, U. A sediment graph model based on SCS-CN method. *J. Hydrol.* **2008**, *349*, 244–255. [CrossRef]
13. Gupta, S.K.; Tyagi, J.; Singh, P.; Sharma, G.; Jethoo, A. Soil Moisture Accounting (SMA) based sediment graph models for small watersheds. *J. Hydrol.* **2019**, *574*, 1129–1151. [CrossRef]
14. Banasik, K.; Krajewski, A.; Sikorska, A.; Hejduk, L. Curve Number Estimation for a Small Urban Catchment from Recorded Rainfall-Runoff Events. *Arch. Environ. Prot.* **2014**, *40*, 75–86. [CrossRef]
15. Kowalik, T.; Walega, A. Estimation of CN Parameter for Small Agricultural Watersheds Using Asymptotic Functions. *Water* **2015**, *7*, 939–955. [CrossRef]

16. Cogliandro, V.; Krajewski, A.; Rutkowska, A.; Porto, P.; Banasik, K. Effect of forest fire on lag time of direct runoff of rainfall in a small catchment in Calabria. *Sylwan* **2017**, *161*, 677–684.

17. Azizian, A.; Shokoohi, A. Development of a new method for estimating SCS curve number using TOPMODEL concept of wetness index (case study: Kasilian and Jong watersheds, Iran). *Acta Geophys.* **2019**, *67*, 1163–1177. [CrossRef]

18. Woodward, D.E.; Hawkins, R.H.; Jiang, R.; Hjelmfelt, J.A.T.; Van Mullem, J.A.; Quan, Q.D. Runoff Curve Number Method: Examination of the Initial Abstraction Ratio. *World Water Environ. Resour. Congr.* **2003**. [CrossRef]

19. Fu, S.; Zhang, G.; Wang, N.; Luo, L. Initial Abstraction Ratio in the SCS-CN Method in the Loess Plateau of China. *Trans. ASABE* **2011**, *54*, 163–169. [CrossRef]

20. Junior, L.C.G.D.V.; Rodrigues, D.B.B.; De Oliveira, P.T.S. Initial abstraction ratio and Curve Number estimation using rainfall and runoff data from a tropical watershed. *RBRH* **2019**, *24*, 24.

21. Lim, K.J.; Engel, B.A.; Muthukrishnan, S.; Harbor, J.; Harbor, J. Effects of Initial Abstraction and Urbanization on Estimated Runoff Using CN Technology. *JAWRA J. Am. Water Resour. Assoc.* **2006**, *42*, 629–643. [CrossRef]

22. Lal, M.; Mishra, S.; Kumar, M. Reverification of antecedent moisture condition dependent runoff curve number formulae using experimental data of Indian watersheds. *Catena* **2019**, *173*, 48–58. [CrossRef]

23. Durán-Barroso, P.; González, J.; Valdés, J.B. Sources of uncertainty in the NRCS CN model: Recognition and solutions. *Hydrol. Process.* **2017**, *31*, 3898–3906. [CrossRef]

24. Dey, P.; Mishra, A. Separating the impacts of climate change and human activities on streamflow: A review of methodologies and critical assumptions. *J. Hydrol.* **2017**, *548*, 278–290. [CrossRef]

25. Zhai, R.; Tao, F. Contributions of climate change and human activities to runoff change in seven typical catchments across China. *Sci. Total. Environ.* **2017**, *605*, 219–229. [CrossRef]

26. Zhou, Y.; Lai, C.; Wang, Z.; Chen, X.; Zeng, Z.; Chen, J.; Bai, X. Quantitative Evaluation of the Impact of Climate Change and Human Activity on Runoff Change in the Dongjiang River Basin, China. *Water* **2018**, *10*, 571. [CrossRef]

27. Chen, Z.; Chen, Y.; Li, B. Quantifying the effects of climate variability and human activities on runoff for Kaidu River Basin in arid region of northwest China. *Theor. Appl. Climatol.* **2013**, *111*, 537–545. [CrossRef]

28. Lv, X.; Zuo, Z.; Xiao, P.; Ni, Y.; Sun, J. Effects of Climate Change and Human Activity on Runoff in a Typical Loess Gullied-Hilly Region Watershed. *Pol. J. Environ. Stud.* **2018**, *27*, 779–785. [CrossRef]

29. Tan, X.; Gan, T.Y. Contribution of human and climate change impacts to changes in streamflow of Canada. *Sci. Rep.* **2015**, *5*, 17767. [CrossRef] [PubMed]

30. Ranzi, R.; Caronna, P.; Tomirotti, M. Impact of Climatic and Land Use Changes on River Flows in the Southern Alps. In *Sustainable Water Resources Planning and Management Under Climate Change*; Kolokytha, E., Oishi, S., Teegavarapu, R., Eds.; Springer: Singapore, 2017; pp. 61–83.

31. Hejduk, L.; Hejduk, A.; Banasik, K. Determination of Curve Number for snowmelt-runoff floods in a small catchment. In Proceedings of the International Association of Hydrological Sciences; IAHS Publ.: Prague, Czech Republic, 2015; Volume 370, pp. 167–170.

32. Banasik, K.; Hejduk, L.; Barszcz, M. Flood flow consequences of land use changes in a small urban catchment of Warsaw. In *11th International Conference on Urban Drainage*; IAHR/IWA: Edinburgh, UK, 2008; pp. 1–10.

33. Sikorska, A.; Banasik, K. Parameter identification of a conceptual rainfall-runoff model for a small urban catchment. *Ann. Warsaw Univ. of Life Sci. SGGW Land Reclam.* **2010**, *42*, 279–293. [CrossRef]

34. Sikorska, A.E.; Scheidegger, A.; Banasik, K.; Rieckermann, J. Considering rating curve uncertainty in water level predictions. *Hydrol. Earth Syst. Sci.* **2013**, *17*, 4415–4427. [CrossRef]

35. Krajewski, A.; Sikorska, A.E.; Banasik, K. Modeling Suspended Sediment Concentration in the Stormwater Outflow from a Small Detention Pond. *J. Environ. Eng.* **2017**, *143*, 05017005. [CrossRef]

36. Krajewski, A.; Banasik, K.; Sikorska, A. Stormflow and suspended sediment routing through a small detention pond with uncertain discharge rating curves. *Hydrol. Res.* **2019**, *50*, 1177–1188. [CrossRef]

37. Krajewski, A.; Wasilewicz, M.; Banasik, K.; Sikorska, A.E. Operation of detention pond in urban area-example of Wyscigi Pond in Warsaw. In *Environmental Engineering V*; Pawlowska, M., Pawlowski, L., Eds.; CRC Press: London, UK, 2017; pp. 211–215.

38. Krajewski, A.; Gladecki, J.; Banasik, K. Transport of suspended sediment during flood events in a small urban catchment. *Acta Sci. Pol. Form. Circumiectus* **2018**, *17*, 119–127. [CrossRef]

39. Majewski, G.; Przewoźniczuk, W.; Kleniewska, M. Precipitation at the meteorological station in Ursynów WULS – SGGW in 1960–2009. *Sci. Rev. Eng. Env. Sci.* **2010**, *19*, 3–22.

40. Majewski, G.; Odorowska, M.; Rozbicka, K. An analysis of the thermal conditions at Ursynów-SGGW station in Warsaw for the years 1970-2009. *Water-Environ.-Rural Areas* **2012**, *2*, 171–184.

41. Department of Water Engineering and Applied Geology (DoWEaAG), Warsaw University of Life Sciences, Warsaw, Poland. *Investigations Realised within the Statutory Activity of the DoWEaAG, Funded by the Ministry of Science and Higher Education, in Years 2010–2017, Internal Report*; Department of Water Engineering and Applied Geology (DoWEaAG), Warsaw University of Life Sciences: Warsaw, Poland, 2017.

42. Copernicus Land Monitoring Service, CORINE Land Cover 2018. Available online: https://www.copernicus.eu/en/services/landURL (accessed on 1 June 2019).

43. Soil-Agricultural map, 1:500,000, Masovian Spatial Information System. Available online: https://msip.wrotamazowsza.pl/msip/Full.aspx (accessed on 1 June 2019).

44. Kurowska-Łazarz, R.; Szulc, W.; Woźniak, B.; Piotrowska, M.; Drożdżyńska, J. *Vademecum. Meteorological MEASUREMENTS and Observations*; IMGW-PIB: Warsaw, Poland, 2015; pp. 9–23.

45. Szumiejko, F.; Wdowikowski, M.; Hański, A.; Kańska, A.; Mielke, M.; Aneszko, J.; Jankowska, I.; Wydrych, M. *Vademecum. Hydrological Measurements and Observations*; IMGW-PIB: Warsaw, Poland, 2015; pp. 13–31.

46. World Meteorological Organisation. *Guide to Hydrological Practice, Hydrology—From Measurement to Hydrological Information*, 6th ed.; World Meteorological Organisation: Geneva, Switzerland, 2008; pp. I.9-12–I.9-14.

47. Kaznowska, E.; Bansik, K.; Hejduk, A.; Krajewski, A.; Wasilewicz, M.; Hejduk, L.; Gładecki, J. Hydrological characteristics since the mid–twentieth century of a small catchment in southern mazovian region, Poland. In *Contemporary Problems of Polish Climate*; Chojnacka-Ożga, L., Lorenc, H., Eds.; IMGW-PIB Press: Warszawa, Poland, 2019; pp. 135–146.

48. Wanielista, M.; Kersten, R.; Eaglin, R. *Hydrology, Water Quantity and Quality Control*; John Wiley&Sons, Inc.: Wiley: New York, NY, USA, 1997; pp. 189–193.

49. Ling, L.; Yusop, Z.; Yap, W.-S.; Tan, W.L.; Chow, M.F.; Ling, J.L. A Calibrated, Watershed-Specific SCS-CN Method: Application to Wangjiaqiao Watershed in the Three Gorges Area, China. *Water* **2020**, *12*, 60. [CrossRef]

50. Walega, A.; Michalec, B.; Cupak, A.; Grzebinoga, M. Comparison of SCS-CN determination methodologies in a heterogeneous catchment. *J. Mt. Sci.* **2015**, *12*, 1084–1094. [CrossRef]

51. Soulis, K.X.; Valiantzas, J.D. SCS-CN parameter determination using rainfall-runoff data in heterogeneous watersheds—the two-CN system approach. *Hydrol. Earth Syst. Sci.* **2012**, *16*, 1001–1015. [CrossRef]

52. Soulis, K.; Valiantzas, J. Identification of the SCS-CN Parameter Spatial Distribution Using Rainfall-Runoff Data in Heterogeneous Watersheds. *Water Resour. Manag.* **2013**, *27*, 1737–1749. [CrossRef]

53. Systat Software Inc. *TableCurve 2D v5.01 for Windows*; Systat Software Inc.: Chicago, IL, USA, 2002.

54. Sikorska, A.; Seibert, J. Value of different precipitation data for flood prediction in an alpine catchment: A Bayesian approach. *J. Hydrol.* **2018**, *556*, 961–971. [CrossRef]

55. Hawkins, R.H.; Hjelmfelt, A.T.; Zevenbergen, A.W. Runoff Probability, Storm Depth, and Curve Numbers. *J. Irrig. Drain. Eng.* **1985**, *111*, 330–340. [CrossRef]

Article

Improvement of SCS-CN Initial Abstraction Coefficient in the Czech Republic: A Study of Five Catchments

Martin Caletka [1,2,*], Monika Šulc Michalková [2], Petr Karásek [3] and Petr Fučík [3]

[1] T. G. Masaryk Water Research Institute, Podbabská 30, 160 00 Praha, Czech Republic
[2] Department of Geography, Faculty of Science, Masaryk University, Kotlářská 2, 611 37 Brno, Czech Republic; sulc@mail.muni.cz
[3] Research Institute for Soil and Water Conservation, Žabovřeská 250, 156 27 Praha, Czech Republic; karasek.petr@vumop.cz (P.K.); fucik.petr@vumop.cz (P.F.)
*   Correspondence: 323980@mail.muni.cz; Tel.: +420-737-651-313

Received: 19 May 2020; Accepted: 3 July 2020; Published: 10 July 2020

Abstract: The SCS-CN method is a globally known procedure used primarily for direct-runoff estimates. It also is integrated in many modelling applications. However, the method was developed in specific geographical conditions, often making its universal applicability problematic. This study aims to determine appropriate values of initial abstraction coefficients λ and curve numbers (*CNs*), based on measured data in five experimental catchments in the Czech Republic, well representing the physiographic conditions in Central Europe, to improve direct-runoff estimates. Captured rainfall-runoff events were split into calibration and validation datasets. The calibration dataset was analysed by applying three approaches: (1) Modifying λ, both discrete and interpolated, using the tabulated *CN* values; (2) event analysis based on accumulated rainfall depth at the moment runoff starts to form; and (3) model fitting, an iterative procedure, to search for a pair of λ, S (*CN*, respectively). To assess individual rainfall characteristics' possible influence, a principal component analysis and cluster analysis were conducted. The results indicate that the CN method in its traditional arrangement is not very applicable in the five experimental catchments and demands corresponding modifications to determine λ and *CN* (or S, respectively). Both λ and *CN* should be viewed as flexible, catchment-dependent (regional) parameters, rather than fixed values. The acquired findings show the need for a systematic yet site-specific revision of the traditional CN method, which may help to improve the accuracy of *CN*-based rainfall-runoff modelling.

Keywords: curve number; direct runoff; HEC-HMS; initial abstraction coefficient; rainfall-runoff modelling; SCS-CN

1. Introduction

The Soil Conservation Service Curve Number (SCS-CN) method is an empirical lumped rainfall-runoff model developed in 1954 by the US Department of Agriculture's Soil Conservation Service (USDA SCS, now the Natural Resources Conservation Service NRCS). Originally, the method entailed converting basic descriptive inputs characterising catchment features into numeric values of curve numbers (*CNs*) [1]. For any given rainfall amount, the corresponding runoff level is estimated. The procedure is based on a combination of the water-balance equation

$$P = Ia + F + Q, \tag{1}$$

and two fundamental assumptions,

$$\frac{Q}{P - Ia} = \frac{F}{S}, \tag{2}$$

$$Ia = \lambda S \tag{3}$$

in which Q is direct runoff (mm); P is precipitation (mm); Ia is initial abstraction (mm) that has to be exceeded so that direct runoff can start to form; F is cumulative infiltration (mm), not including Ia; S is maximum potential retention of a watershed after the start of the runoff and λ is initial abstraction ratio (dimensionless).

The runoff depth Q is expressed using Equations (1)–(3) as

$$Q = \frac{(P - \lambda S)^2}{(P + (1 - \lambda)S} \text{ for } P > Ia, \ Q = 0 \text{ for } P \leq Ia. \tag{4}$$

The maximum potential retention S is expressed as a function

$$S = \frac{25400}{CN} - 254, \tag{5}$$

in which CN is the curve number, theoretically ranging between 0 and 100. Over the years, the method has undergone numerous modifications [2–6], during which time the scientific community adopted the method globally for its simplicity, applicability and relatively low demands on input data [7–11].

The CN method has various shortcomings regarding practical applications in different geographical conditions. For example, the initial abstraction coefficient λ's validity has been questioned [11–14]. Similarly, the role of a watershed's relief and land cover [15–20], previous rainfall's influence [19–21] and CN values' representativeness [16,22,23] in terms of properly setting the method have been scrutinised.

Attention has been payed to the ability of a single CN value to characterise the runoff response of a watershed correctly. While the CN may be calculated from observed P-Q data, practical experience shows that the CN values vary between events [11,24]. Therefore, various approaches were introduced aiming to estimate an appropriate single CN value or single asymptotic CN, separately [1,25–29]. However, the aforementioned methods showed certain shortcomings addressed by a study that analysed the effect of soil and land cover complexes on the hydrologic responses in watersheds [30]. Specifically, the authors proposed a concept of a two-CN heterogenous system. This study was subsequently generalised [31]. The latter two methods along with other CN estimation methods were used and compared in a study [32] of a watershed affected by a major forest fire, so that the effect of both spatial heterogeneity in land cover and the fire could be investigated.

Another problematic aspect of the CN method lies in the characterisation of previous rainfall. It is expressed using three antecedent moisture condition (AMC) classes (I: Dry; II: Normal; and III: Wet), each of which has its own set of CNs. The discrete character of the CN-AMC relationship is manifested by sudden jumps in estimated runoff [33]. On the basis of previous five-day rainfall totals, CN_{II} is converted to CN_I and CN_{III} using relationships proposed by [24–37] and directly from the NEH-4 tables [4,38,39]. Despite the existence of different conversion methods, the discontinuity problem remained unresolved, and the concept of soil moisture accounting (SMA) became a subject of research. Williams and LaSeur [40] presented the first SMA concept. Their procedure was based on the idea that the CN varies continuously, along with soil moisture. Hawkins [41] then developed a continuous soil moisture accounting concept, but this modification did not take into account the SCS-CN procedure's structural shortcomings in terms of SMA, a problem addressed by Mishra et al. [42], who introduced an improved SMA-based SCS-CN model. Subsequently, Mishra and Singh [43] developed a versatile SCS-CN model (VSCS-CN) based on the assumption that higher antecedent soil moisture before a rainfall event will provide greater runoff and lower the soil's capacity to store and retain rainfall water.

Similarly, Michel et al. [44] also examined such a concept. Mishra and Singh [21] used a concept that comprised a runoff coefficient and the degree of saturation on the basis of which multiple improved SCS-CN models have been proposed [19,45–49].

Relief is a significant factor in runoff formation [50,51], but slope was not reflected in the traditional CN model. The NRCS handbook's CN values [6] are viewed as corresponding to a 5% slope [18,20,51,52]. However, some approaches exist that incorporate slope into the CN estimates. For example, Huang et al. [51] modified the procedure by Sharpley and Williams [18] for steep watersheds in Loess Plateau, China. Their method then was adjusted for rainfall-runoff data from 39 South Korean watersheds. However, according to Mishra et al. [8], the aforementioned procedures require proper validation.

Inaccuracies in direct-runoff modelling are frequently associated with the initial abstraction Ia or λ, separately. Ia was not incorporated into the original CN method [52]. During its evolution, a constant value of 0.2 was set for λ [4,53], but many authors challenged this. Trying to identify this constant's appropriate value has been a matter of many debates and studies conducted to estimate the appropriate value of λ. It was concluded by several studies that λ values result from multiple factors related to rainfall events and landscape features [22,54,55] and that the coefficient should be viewed as a regional parameter expressing geographical variability [10]. Some researchers concluded stated that λ can take any non-negative value, in some cases even greater than 1 [55–59]. However, many authors stated that using reduced λ led to much more accurate results, e.g., [2,11–13,60–64].

In the Czech Republic, the CN method is applied frequently, using tabulated *CN* values [65]. The value of λ implicitly is set at 0.2, but such a value does not appear to be representative, as preliminary studies by Caletka and Honek [66] and Caletka and Michalková [67] concluded. The CN method is the subject of greater interest in Slovakia where several studies were conducted in recent years. Karabová [68] analysed rainfall-runoff data from five lowland watersheds in Slovakia, addressing the problem of regionalisation of runoff curve through a new AMC classification and application of λ–P regressions. The study concluded that the traditional CN method lead to over-estimated direct runoff for lower P and vice versa. Similarly, Kohnová et al. [69] analysed rainfall-runoff data from nine watersheds in central Slovakia and concluded that the standard CN method demanded certain modifications. Specifically, they stated that *CN*s should be estimated on the basis of observed rainfall-runoff data, rather than using tabulated values. Moreover, they used the variable λ value instead of the fixed one. Another study [23] compared the original CN method's performance with that based on empirical rainfall-runoff events using different approaches. They stated that the CN method was burdened by a lot of uncertainty. Using asymptotic fitting procedure of median procedure with reduced $\lambda = 0.05$ increased the accuracy of the results. Recently, Kohnová et al. [70] proposed and evaluated the performance of a regionally based approach to estimate *CN* along with different λ values. They collected data in a group of Slovak and Polish catchments and used the *L*-moment based method and ANOVA for delineation of two homogenous regions of catchments with similar *CN*s. The optimal value of λ was set to 0.15.

Given that the application of the CN method in its original form is problematic, based on findings in many studies, this research focuses on an analysis of recorded rainfall-runoff data from five experimental watersheds in the Czech Republic with various geomorphological and agricultural conditions. This study's main objectives are: (1) To calculate adjusted λ both as discrete values and regression functions; (2) to compare the CN method's performance using adjusted λ values; and (3) to test the application of the modified CN method in HEC-HMS simulations of specific rainfall-runoff events in a chosen catchment.

2. Study Area

The study was carried out in five experimental watersheds in the Czech Republic (Figure 1), differing mainly in size and in land use. This made it possible to assess the CN method's performance and interpret the findings of this research with regard to each watershed's specific geographical conditions. All the studied watersheds' basic characteristics are provided in Table 1.

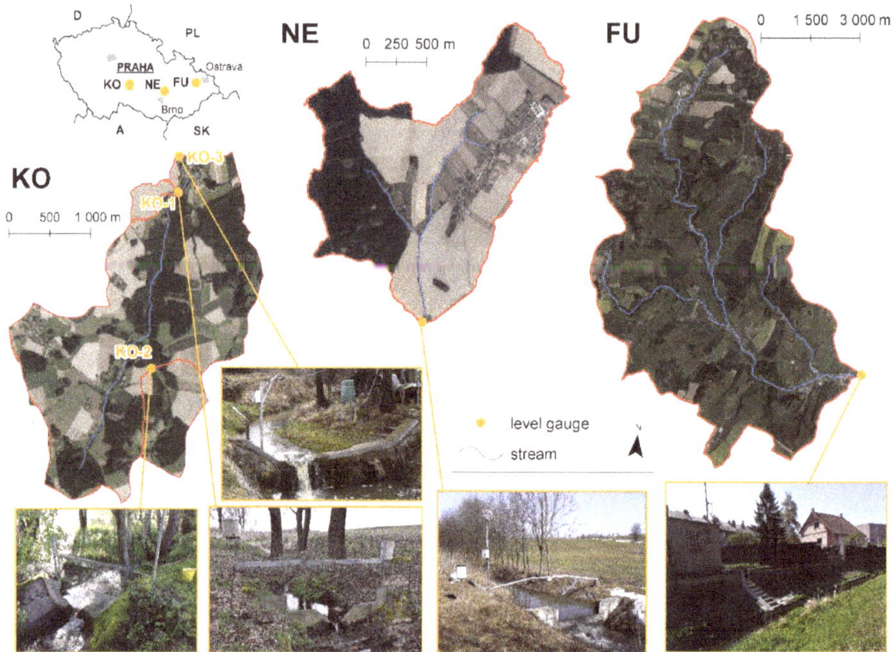

Figure 1. Overview map of the experimental catchments.

Table 1. Basic characteristics of the experimental catchments and percentage of land use categories (AL—arable land; BG—bare ground; FO—forest; GR—grassland and meadows; GA—gardens and orchards; PA—paved and built-up areas; SH—shrubbery; WA—water bodies).

Sub-Catchment	Area (km^2)	Elevation (m)		Mean Slope (°)	Mean P (mm)		Mean Q (m^3/s)	Mean T (°C)	
		Min.	Max.		Annual	IV-IX		Annual	IV-IX
FU	57.8	282	563	6.1	675	450	0.350	8.0	13.5
NE	2.8	559	650	4.2	650	400	0.015	7.0	14.0
KO-1	0.16	490	532	3.2	675	425	0.007	6.5	12.5
KO-2	0.78	548	623	4.3	675	425	0.003	6.5	12.5
KO-3	7.1	478	623	5.2	675	425	0.026	6.5	12.5

Sub-Catchment	Percentage of Land Use Category (%)								
	AL	BG	FO	GR	GA	GA	PA	SH	WA
FU	43.5	0.0	24.7	23.7	5.2	5.2	1.4	1.3	0.2
NE	50.9	0.2	31.8	5.2	5.4	5.4	3.1	3.4	0.0
KO-1	97.3	0.0	0.1	0.0	1.3	1.3	0.7	0.6	0.0
KO-2	52.2	0.0	44.4	1.1	0.6	0.6	0.1	1.5	0.0
KO-3	41.8	0.0	37.9	8.0	6.1	6.1	0.9	5.2	0.1

2.1. Husí Creek Catchment (FU)

Husí Creek is a left-side tributary of the Oder (Odra) River, located in northeastern Czech Republic, approximately 30 km southwest of the city of Ostrava. The upper parts belong to the easternmost edge of the Bohemian Massif in the Nízký Jeseník Highlands, while the lower parts lie in the Outer Subcarpathia, a depression at the outer base of the Carpathian arc. The catchment's position predetermines its specific relief; it is characterised by a flat or slightly wavy surface in the upper parts that gradually passes into narrowing valleys with greater longitudinal slope. The valleys widen downstream, and the surface becomes slightly curved to flat. Two main soil types are present in the catchment. Cambisols are found mainly in the upper parts, while luvisols are prevalent in the lower parts. As for land use, a specific spatial distribution exists. Arable land, usually in the form of large plots, is found both in the upper and lower parts of the catchment on the flat relief. Forests and meadows are located mostly in the transition between the upper and lower parts, in the valleys and along streams. This specific environment creates favourable conditions for generating flash floods from intense rainfall. Two major flash-flood events were recorded in summer 2009 and 2010 [71,72]. On 2 July 2009, the region experienced extreme rainfall (local precipitation reached up to 50 mm) after a very rainy period, causing a severe flood event with peak discharge of 33.1 m^3/s and corresponding to a 50-year flood event. On 1 June 2010, another rainfall event was recorded (with a maximum rain-gauge measurement of 55 mm), but due to the precipitation's differing spatial distributions and lower previous rainfall, peak discharge reached only 14.4 m^3/s, corresponding to the return period of 5–10 years. Nevertheless, both events resulted in heavy damage to municipalities along the stream.

2.2. Suchý Creek Catchment (NE)

The study area is located approximately 30 km north of Brno in the Drahany Highlands. The catchment is drained by a minor stream, Suchý Creek, and forests cover the western part of the watershed. More than half the area is arable land. Due to large fields, soil erosion poses an imminent threat to a little dam that was constructed right under the catchment's outlet to protect the areas downstream from flash floods [73].

2.3. Kopaninský Creek Catchment (KO)

This catchment is located in the Křemešnice Highlands in the southern part of the Sázava River basin, a right-side tributary of the Moldau (Vltava) River. It is a source of drinking water for the City of Prague, located approximately 80 km to the northwest. From a geological perspective, this is a very old area formed by igneous and metamorphic rocks during the Paleozoic Era, including granite and schist. Rolling hills and low mountains dominate this region's surface, with the altitude mostly ranging between 500 and 650 m. As for the soils, cambisols are found most frequently here, along with pseudogleys and gleys, in valleys and along streams. The landscape is covered predominantly by arable land, forests and, to a lesser extent, grasslands and meadows. The settlement is scattered in rather small villages.

The KO catchment is drained by the Kopaninský Creek. The main catchment outlet is KO-3, another two sub-catchments analysed in this study are KO-2 and KO-1. The KO-2 catchment is located in the uppermost part of the KO-3 catchment and has the greatest portion of forested areas, at more than 44%. The KO-1 catchment, located in the lower part, is covered almost entirely by arable land (more than 97%) and is partly tile-drained.

All five studied watersheds' basic geographical characteristics are listed in Table 1.

3. Materials and Methods

3.1. Data Collection

Automatic rain gauges were used to measure precipitation heights. The monitoring network included up to 17 rain gauges (automatic 0.2 mm tipping buckets 200 cm^2 and 0.1 tipping buckets 500 cm^2) in catchment FU and its nearby surroundings. The Thiessen polygon interpolation procedure was used to calculate average precipitation levels [74]. A single rain gauge (automatic 0.2 mm tipping bucket 200 cm^2) was used to measure rainfall in catchments NE and KO. Using the precipitation data, individual rainfall events were identified, for which five- and 10-day antecedent precipitation totals (P_{5d}, P_{10d}) were calculated.

The water level data were measured using calibrated weirs with ultrasound probes (Fiedler company)—a Thompson weir was used with in catchments NE, KO-1 and KO-2, a compound Cipolletti weir was used in catchment KO-3. In catchment FU, water levels were measured using hydrostatic level sensors (Fiedler company). Discharge values were calculated by means of rating curves set for all monitored outlets on the basis of hydrometric measurements, levelling and using the Chézy formula [75].

An overview of measuring devices is provided in Table 2.

Table 2. Overview of rainfall and level/runoff measurements in the study catchments; Number of events chosen for training and validating datasets.

Sub-Catchment	Period	Number of Rain Gauges (Interval)	Level/Runoff Measurement (Interval)	Number of Events Chosen (Train./Valid.)
FU	2008–2016	17 (10 min)	Level gauge (1 h)	44/45
NE	2008–2017	1 (10 min)	Thomson weir (10 min)	57/81
KO-1	2005–2018	1 (10 min)	Thomson weir (10 min)	56/52
KO-2	2005–2018	1 (10 min)	Thomson weir (10 min)	73/74
KO-3	2005–2018	1 (10 min)	Cippoletti weir (10 min)	80/60

3.2. Estimate of CN Values

The hydrologic soil groups (HSGs) were derived based on Valuated Soil-Ecological Units (VSEUs), a whole-country database (1:5000) determined by the Research Institute for Soil and Water Conservation (RISWC). Land-cover units were identified based on aerial images and a reconnaissance field survey. The corresponding percentages are provided in Table 1, and the spatial distribution is depicted in Figure 2. Moreover, cropping and tillage history also was considered in the NE catchment. Subsequently, the final identification of *CN*s for each unit was done according to Janeček et al.'s [56] methodology, which frequently is applied in the Czech Republic, in which the tabulated *CN* values correspond to different HSGs, land-use units and crop types. Moreover, moisture conditions are reflected through antecedent moisture conditions (AMCs). AMCs are characterised by three degrees (I: Dry; II: Normal; and III: Wet), delimited by a five-day precipitation total of P_{5d} of 36 mm and 53 mm prior to a rainfall ($P_{5d} < 36$ mm—dry conditions; $P_{5d} \geq 53$ mm—wet conditions). Maps of CN_{II} values are provided in Figure 2 for each catchment.

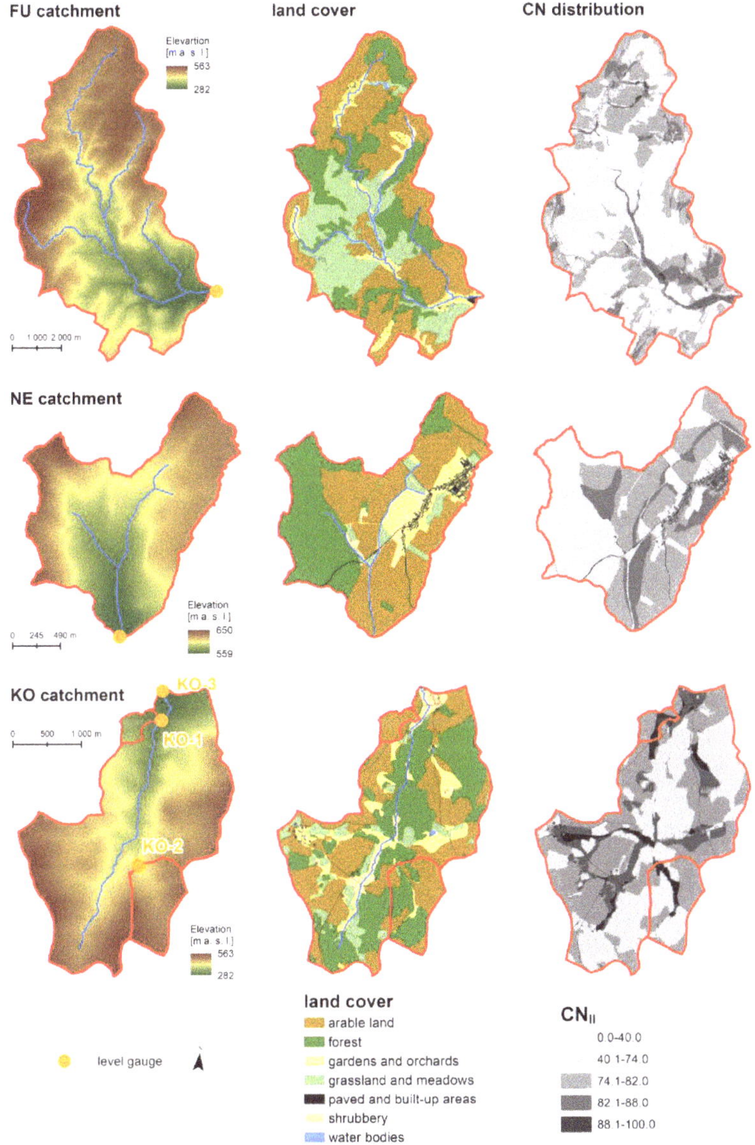

Figure 2. Surface, land cover classification and curve number (*CN*) determination.

3.3. Events, Determination of λ

To avoid biased λ values, rainfall events recorded between April and September with $P \geq 5$ mm were chosen for the analysis. In order to exclude rainfall events that caused compound hydrographs, rainfall-runoff events were selected individually, so that the influence of runoff from previous rainfall events could be obviated. For each event, the following parameters were determined:

1. precipitation level P (mm),
2. rainfall event's duration (h),
3. maximum rain intensity i_{max} (mm h^{-1}),
4. five- and 10-day rainfall accumulations P_{5d} and P_{10d} (mm),
5. hydrograph was drawn and base flow was separated from the hydrograph based on the hydrograph's inflection points, which indicated the start of the rising limb and the end of the recession limb for each runoff event [69,70],
6. direct-runoff level Q was calculated (mm),
7. initial abstraction Ia (mm), which was calculated as the rainfall level at the moment when direct runoff begins. Such a procedure has been used for instance in research by Shi et al. [11]. However, in large watersheds and/or in case of uneven spatial distribution of precipitation such a calculation of Ia might be problematic as the runoff needs some time to reach the watershed's outlet [24].

The aforementioned parameters were applied in the analyses described in detail in the following sections. To verify the adjusted CN method's performance, events' datasets were split into training and validating datasets (see the numbers provided in Table 2), so that they were similar in terms of level and variability of P, P_{5d} and P_{10d}.

3.3.1. Principal Component Analysis, Cluster Analysis

In order to reveal possible concealed relationships between parameters of rainfall-runoff events and to assess objectively the role that the individual parameters may play in the rainfall-runoff process, principal component analysis (PCA) and cluster analysis (Ward's method) were conducted for the training datasets from all experimental catchments. The cluster analysis allowed to explore groups (clusters), which the events tend to form. The optimal number of clusters was set using Pham et al.'s [76] k-selection method.

3.3.2. Modified λ with Tabulated CNs

The procedure begins with the determination of tabulated CN values [65] for individual units in a catchment following the procedure described in Section 3.2. For each unit, from the CN values, maximal potential retention S values are derived using Equation (5). Average S for the entire watershed area is calculated using weighted arithmetic mean. In the following step, the corresponding initial abstraction coefficient is calculated using Equation (4), which is converted into the quadratic formula

$$\left(-S^2\right)\lambda^2 + (2PS - QS)\lambda + \left(QP + QS - P^2\right) = 0, \tag{6}$$

producing a pair of initial abstraction coefficient values for each event. From these, the final selection is carried out following the assumption that the lowest nonnegative value must be chosen. Numerically, the solution of the quadratic equation is correct. However, initial abstraction coefficient λ must be higher than or equal to zero. In case of two non-negative roots, choosing the higher one causes initial abstraction Ia being too high, resulting in zero direct runoff generated as the condition from Equation (4) is not met.

The determination of S values and subsequent calculation (Equation [6]) were conducted twice: (1) Taking into consideration the AMC classes (i.e., using CN_I, CN_{II} and CN_{III}); and (2) neglecting AMC classes and using only CN_{II}, assuming that the AMC determination might not be a deciding factor in calculating direct runoff.

Discrete λ

This procedure's objective was to identify a discrete representative λ for a watershed calculated as mean and median of values obtained using Equation (6). Mean and median λ were calculated for the entire training dataset of events and for clusters delimited by the cluster analysis.

Interpolated λ

This procedure's objective was to express λ as a linear function of P [66], instead of using a single value. The regression equations were derived using λ:P plots. Again, the calculations were carried out with and without determination of AMC.

3.3.3. λ Modifications Not Dependent on Tabulated CNs

Event Analysis

This procedure was an inverted process. The maximum potential retention, S, and λ were calculated on the basis of rainfall and runoff parameters. The calculation included the following steps:

1. The total runoff Q and the initial abstraction Ia were calculated using the procedures described above (Section 3.3).
2. The maximum potential retention, S, was an unknown parameter that was calculated using equation

$$S = \frac{(P - Ia)^2}{Q} - (P - Ia),\tag{7}$$

obtained by modifying Equation (4).

3. The initial abstraction coefficient λ values were determined by dividing Ia by S for each rainfall-runoff event. Mean and median λ were calculated for each experimental watershed.
4. Regression of S, according to P_{10d}, was calculated and used for the validation together with mean and median λ.

Model Fitting

Model fitting is an iterative least-squares procedure for determining a pair of λ and S such that the sum of the squared differences between the runoff observed and modelled

$$\sum \left[Q - \frac{(P - \lambda S)^2}{(P + (1 - \lambda)S)} \right]^2 \tag{8}$$

as a minimum. The computation of the model-fitting procedure was carried out using a loop-based script in Python 3.7.

3.4. Comparison of Estimated Runoff

For each experimental catchment, a training dataset of events was chosen to identify the key relationships characterising the CN method's proper setting. The remaining events were used for the parameter validation.

The quality of the runoff levels estimated by the modified CN method was evaluated by means of Nash–Sutcliffe model efficiency coefficient (*NSE*) (9), and bias (*e*) (10).

NSE is given as

$$NSE = 1 - \frac{\sum \left(Q_{pi} - Q_i \right)^2}{\sum \left(Q_i - Q_a \right)^2},\tag{9}$$

in which Q_{pi} is the *i*-th modelled runoff depth, Q_i is the *i*-th observed runoff depth and Q_a is the mean observed runoff depth.

Bias (*e*) is given as:

$$e = \frac{\sum_{i=1}^{n}\left(Q_{pi} - Q_i\right)}{n}. \tag{10}$$

NSE can range from $-\infty$ to 1. An *NSE* equal to 1 expresses a perfect match of the modelled and observed values. An *NSE* ≥ 0.5 indicates a very good prediction ability for the model. A lower *NSE* indicates a worse performance by the simulations. The *e* is the average difference between the modelled and predicted runoff, thereby expressing the model's under- or over-estimating tendencies.

3.5. HEC-HMS Simulations

To test the applicability of the modified SCS-CN parameters for simulation of surface runoff, HEC-HMS 4.3 software was used. The simulations' aim was to verify the model's ability to deliver a runoff volume/level that did not differ from the observed one by more than 10%, with an *NSE* ≥ 0.500 (i.e., corresponding to the observed hydrograph's time course). Such simulations are viewed as sufficiently accurate. The computations were carried out only in the NE catchment, which is relatively small (A = 2.8 km^2), thereby allowing for better simulation of the surface runoff and identifying the surface-runoff simulations' response to changes in individual parameters provided in Table 3.

Table 3. Overview of methods employed in surface runoff simulations using HEC-HMS.

	Method	Parameter
Loss	SCS Curve Number	Initial abstraction (mm) Curve Number (-) Impervious (%)
Transform	Clark Unit Hydrograph	Time of concentration (hr) Storage coefficient (hr)
Baseflow		Initial discharge (m^3/s) Recession constant (-) Ratio (-)

The catchment was split into three sub-basins. Calibration was carried out for two types of runoff events: (1) Those with a gradual onset, and (2) those with both a rapid onset and rapid recession limb.

The initial abstraction, *Ia*, was calculated using the curve numbers estimated for the sub-basins through the method described in Section 3.3, the λ values from the original SCS-CN method (λ = 0.2) and the median λ values (λ = 0.0142) from the analysis of which the results are provided in Section 4.2 The percentage of impervious surfaces (buildings and sealed roads) was calculated using ESRI ArcGIS 10.7 software. Time of concentration, T_c, was calculated using Dingman's [77] formula,

$$T_c = 1.67 \, \frac{3.281 \, L^{0.8}\left(0.0394 \, A + 1\right)^{0.7}}{1900 \, \sqrt{Y}}, \tag{11}$$

in which L is the longest flow distance path in a sub-basin (m), A is the sub-basin's area (km^2) and Y is mean slope (%). Dingman's formula (11) commonly is used in calculations by the Czech Hydrometeorological Institute (CHMI).

The initial discharge at each sub-basin's outlet was calculated using the main outlet's value according to the percentage of the sub-basin areas in the entire NE catchment. The remaining parameters (i.e., storage coefficient, recession constant and ratio) were set so that the calibration delivered the best possible result expressed by the bias of the runoff volume/level and *NSE*. Calibration coefficients were calculated for *Ia*, *CN* and T_c as rates of the calibrated values to the original ones.

Finally, two sets of these values were available: (1) For the runoff events with slow onset, and (2) for runoff events with rapid onset.

In the subsequent step, two verification events were chosen—one with a slower rising limb and another with rapid onset. Having known the P_{5d} (*AMC*, respectively), the original *CN*, *Ia* and T_c values were used for both $\lambda = 0.2$ and $\lambda = 0.0142$, corresponding to each verification event's year. These values were multiplied by the calibration coefficients acquired in the calibration simulations. The storage coefficient, the recession constant and the ratio were calculated in the calibrations.

4. Results

4.1. Influencing Parameters

4.1.1. Principal Component Analysis

To reveal various rainfall and runoff parameters' influence on runoff response to rainfall at event level, principal component analysis (PCA) and cluster analysis were conducted for the studied watersheds.

The PCA analyses' outputs are presented in Figure 3. Table 4 provides an overview of the variability percentages that the first two principal components (PC1 and PC2) explain, including all input parameters' loading values in relation to PC1 and PC2. The variability percentages explained by PC1 and PC2 differ between the studied catchments. In the FU catchment, PC1 showed a value of over 71% and PC2 less than 14%. In the other catchments, PC1 and PC2 percentage differences were not as big. The least difference was recorded in the KO-1 catchment (PC1 49%, PC2 40%). The vectors' orientation in Figure 3 and the loading values in Table 4 indicate that P_{10d} and P_{5d} present the most influential parameters in the direction of PC1, with maximal rain intensity and P being the most influential parameters in the direction of PC2. The KO-1 catchment has an opposite arrangement, as maximal rain intensity and P have the highest loading values in PC1 and P_{10d} and P_{5d} in PC2. This might be caused by the high percentage of arable land with presence of tile drainage system [78]. In such environment, P_{10d} and P_{5d} do not play the most influential role. Instead, direct runoff formation is dominated by the rain intensity along with P. Higher rain intensity produces faster and higher surface runoff and vice versa. This applies in general, but is probably accentuated in this specific catchment, making it different from the other ones.

To sum up, PCA indicates that P_{10d} and P_{5d} are the most influential parameters. The effects from maximal rain intensity and P, the other two most influential parameters, can be viewed as opposites, as they act along PC2.

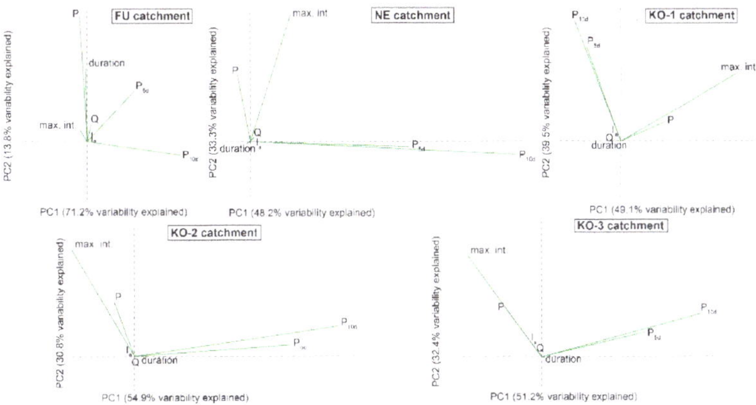

Figure 3. Principal component analysis PCA conducted for rainfall-runoff events in the experimental catchments.

Table 4. Quantitative outputs of the principal component analysis (PCA) for each study catchment: Percentage of variability explained by PC1 and PC2 along with the loading values for each input parameter.

| Sub-Catchment | FU | | NE | | KO-1 | | KO-2 | | KO-3 | |
| Explained | PC 1 | PC 2 | PC 1 | PC 2 | PC 1 | PC 2 | PC 1 | PC 2 | PC 1 | PC 2 |
Variability	71.2%	13.8%	48.3%	33.3%	49.1%	39.5%	54.9%	30.8%	51.2%	32.4%
P	−0.07	0.81	−0.04	0.46	0.32	0.11	−0.08	0.44	−0.18	0.41
duration	−0.02	0.47	−0.01	−0.02	−0.02	0.00	0.02	0.00	0.02	−0.01
P_{5d}	0.45	0.32	0.51	−0.04	−0.25	0.56	0.59	0.09	0.50	0.20
P_{10d}	0.89	−0.09	0.85	−0.09	−0.34	0.71	0.77	0.24	0.77	0.34
max. int.	−0.07	0.07	0.13	0.88	0.85	0.40	−0.24	0.86	−0.35	0.81
Q	0.04	0.11	0.01	0.03	0.00	0.00	0.00	0.01	−0.01	0.04
Ia	−0.01	0.01	0.00	0.02	0.00	0.01	−0.01	0.02	−0.05	0.11

4.1.2. Cluster Analysis

Using the methodology introduced by Pham et al. (2005), the optimal number of clusters was determined to be two for the experimental watersheds. Subsequently, non-hierarchical k-means clustering was conducted. Mean and median values of variables in the delimited clusters are listed in Table 5. A difference in P_{5d} and P_{10d} between the delimited clusters in the studied watersheds is apparent. Such a finding corresponds to the output of the cluster analysis. Given that P_{10d} was identified as the most important factor, P_{10d} threshold values were calculated for the catchments to be able to distinguish the cluster to which an event belongs. To determine thresholds, the mean of the upper quartile P_{10d} from the cluster with smaller P_{10d} and the lower quartile P_{10d} from the other cluster (i.e., a cluster with higher P_{10d}) was calculated. The same procedure was applied to calculate the corresponding P_{5d} thresholds, as the P_{5d} normally is used in the traditional SCS-CN procedure to determine AMC classes. The P_{5d} and P_{10d} thresholds were 26 mm and 47 mm in the FU catchment, 18 mm and 39 mm in the NE catchment, 35 mm and 42 mm in the KO-1 catchment, 41 mm and 50 mm in the KO-2 catchment and 20 mm and 28 mm in the KO-3 catchment. For each catchment, these P_{5d} and P_{10d} thresholds were used for separation of rainfall-runoff events into two groups in analyses across the paper.

Table 5. Mean and median values of events' parameters (duration (h), P (mm), P_{5d} (mm), P_{10d} (mm), max. int. (mm/h), Q (mm), *Ia* (mm)) in the delimited clusters for the experimental catchments.

| | | FU | | | NE | | | KO-1 | | | KO-2 | | | KO-3 | |
		Mean	Med.		Mean	Med.		Mean	Med.		Mean	Med.		Mean	Med.
duration		12.02	10.00		3.62	2.67		4.00	2.92		3.50	2.67		3.47	2.75
P		17.4	14.0		10.8	6.9		17.1	11.5		13.9	10.0		14.3	10.0
P_{5d}	all (n = 45)	25.7	20.1	all (n = 81)	16.6	11.3	all (n = 56)	18.4	11.9	all (n = 73)	14.7	9.2	all (n = 80)	14.4	8.7
P_{10d}		42.9	35.3		30.1	22.2		28.8	21.2		25.0	18.8		27.2	19.9
max. int.		10.68	8.30		14.40	7.80		23.12	13.80		19.11	12.60		19.88	12.30
Q		1.830	0.337		0.448	0.076		0.316	0.081		0.293	0.069		0.567	0.101
Ia		1.5	0.8		1.7	1.4		2.4	1.8		0.8	0.0		2.4	1.1
duration		12.03	9.00		3.81	2.92		3.69	2.92		3.41	2.80		3.37	2.67
P		18.5	14.1		11.3	7.1		17.8	11.4		14.0	10.0		16.3	10.8
P_{5d}	cl. 1 (n = 31)	17.1	13.3	cl. 1 (n = 58)	8.3	6.7	cl. 1 (n = 48)	11.8	10.2	cl. 1 (n = 68)	11.2	8.5	cl. 1 (n = 52)	7.0	5.4
P_{10d}		26.2	26.4		18.1	17.0		22.4	18.8		21.8	17.2		13.7	13.7
max. int.		11.43	8.90		13.70	7.80		24.40	14.40		19.43	12.90		22.80	12.60
Q		1.152	0.301		0.319	0.065		0.308	0.056		0.246	0.052		0.633	0.090
Ia		1.6	0.8		1.7	1.5		2.4	1.6		0.8	0.0		2.9	1.2
duration		12.00	10.50		3.15	2.33		5.81	5.17		4.67	2.67		3.67	2.83
P		15.1	12.4		9.7	5.5		12.9	11.7		12.5	10.8		10.5	8.8
P_{5d}	cl. 2 (n = 14)	44.7	42.7	cl. 2 (n = 23)	37.7	37.4	cl. 2 (n = 8)	58.2	60.3	cl. 2 (n = 5)	63.2	63.2	cl. 2 (n = 28)	28.2	25.3
P_{10d}		79.9	76.6		60.4	59.5		67.5	67.4		68.0	69.9		52.3	42.5
max. int.		9.03	7.00		16.17	9.00		15.60	7.80		14.90	10.80		14.55	11.40
Q		3.332	2.406		0.773	0.097		0.361	0.208		0.925	0.514		0.444	0.178
Ia		1.3	1.0		1.6	1.3		2.6	2.4		0.8	0.2		1.4	0.7

4.2. Tabulated CN-Based λ Modification

The first way to improve the CN method is through incorporation of tabulated *CNs* [65] along with modified λ values. The following calculations were carried out with and without taking into consideration the separation of rainfall events according to the P_{5d} and P_{10d} thresholds determined through the cluster analysis. Moreover, AMC classes' effect was examined by performing the calculations with and without the AMC classification. Modified λ was calculated as a discrete value and as functionally dependent on P.

4.2.1. Discrete λ

For the comparison, the original CN method with $\lambda = 0.2$ was applied, and the results showed that in this form, the model mostly was unable to deliver good results, except for the KO-1 catchment (for $\lambda = 0.2$ $NSE = 0.861$). Partial improvements were recorded in simulations with the AMC classification omitted. Only in the KO-1 catchment, the accuracy reduced significantly (NSE dropped from 0.861 to -16.769, see Table 6 and Figure S1a,b). It is important to emphasize that the traditional CN method is limited in terms of "zero runoff" cases since the condition from Equation (4) is often not met due to high λ, and also due to the effect of spatial variability in catchments [30,31], which, however, was not addressed in the present study. The relatively good results for the traditional CN method needs to be confronted with high numbers of events for which the model was unable to deliver non-zero direct runoff. This is apparent in Figure S1a,b. Specifically, the number of non-zero cases ranged from 0 to 13 for simulations with AMC classified and from 7 to 27 for simulations with AMC implicitly set as normal (AMC = II).

To improve the CN method's performance, mean and median λ values were determined in the training datasets as a whole and in groups separated according to the P_{5d} and P_{10d} thresholds. Generally, the analysis revealed that in most instances λ was lower than 0.2. In the unseparated datasets, mean λ ranged from 0.0365 to 0.1623 and median λ from 0.0142 to 0.0913. Except for one catchment (KO-1), the reduction of λ led to improvements in estimated runoff, since NSE exceeded 0.500, thereby indicating sufficient accuracy. In simulations with separation of events according to P_{5d} and P_{10d} thresholds, a greater variability in λ values existed, ranging from 0.0183 to 0.4231. Despite the expectation that this greater diversification of λ values will lead to more accurate results (in relation to the non-cluster-based approach—all, i.e., without the separation according to P_{5d} and P_{10d}), this assumption only partially was fulfilled. In the FU catchment, the cluster-based approach improved all direct runoff estimates. In the NE catchment, better results were delivered for median. In catchments KO-2 and KO-3, only simulations with median λ applied in events separated according to P_{10d} led to increased accuracy in estimates. No positive response was recorded in simulations using the modified λ values in the KO-1 catchment (NSE ranged from -7.993 to 0.323).

In examining the calculations while not considering AMC classes (AMC implicitly was set to II, normal), higher accuracy in calculated runoff was recorded in the original CN approach, except for the KO-1 catchment. The cluster-based simulations led to better runoff matches only in the FU and NE catchments. In the other catchments, accuracy even decreased significantly.

In summary, the results show that replacing λ originally set to 0.2 with lower values can lead to improvements in direct-runoff simulations. However, the KO-1 catchment was the only one in which no positive effect from the λ modification was recorded. High e values indicated the model's over-estimating tendency in this specific catchment, probably due to improper CN determination. This might have been caused by the high percentage (97.3%) of arable land along with the presence of the tile drainage systems [78]. In other words, the results revealed quite specific hydrologic conditions in KO-1 catchment compared to the other two KO catchments. Moreover, this well illustrates that CN determination based solely on the prescribed criteria [65] might be misleading. Instead, it should take into consideration all available data on an area of interest.

It should be noted that some studies stated that the modification of λ should be accompanied by a corresponding adjustment in CN or S, separately [2,16]. Therefore, such an approach also was used in this research, but this did not lead to better results than those presented above.

Table 6. Accuracy of direct runoff simulations using mean and median λ values in the experimental catchments with (antecedent moisture condition (AMC) all) and without (AMC = II) determination of AMC classes. $NSE \geq 0.500$ indicating sufficient accuracy are highlighted.

					AMC All			AMC = II		
			λ		r^2	NSE	e	r^2	NSE	e
FU		original	$\lambda = 0.2$		0.258	0.152	−0.818	0.854	0.361	−0.876
	median	all			0.559	0.448	−0.123	0.677	0.658	+0.148
		P_{5d}	≤26 mm	>26 mm	0.612	0.573	−0.115	0.673	0.655	+0.148
		P_{10d}	≤47 mm	>47 mm	0.621	0.587	−0.084	0.674	0.655	+0.141
	mean	all			0.413	0.282	−0.591	0.763	0.663	−0.489
		P_{5d}	≤26 mm	>26 mm	0.563	0.516	−0.532	0.671	0.627	−0.400
		P_{10d}	≤47 mm	>47 mm	0.57	0.527	−0.512	0.673	0.627	−0.404
NE		original	$\lambda = 0.2$		0.01	0.071	0.298	0.781	0.388	0.253
	median	all			0.671	0.668	+0.052	0.684	0.465	+0.173
		P_{5d}	≤39 mm	>39 mm	0.676	0.676	+0.017	0.675	0.436	+0.192
		P_{10d}	≤39 mm	>39 mm	0.68	0.68	+0.012	0.677	0.442	+0.188
	mean	all			0.667	0.619	−0.103	0.717	0.698	−0.01
		P_{5d}	≤39 mm	>39 mm	0.617	0.59	−0.017	0.697	0.671	+0.011
		P_{10d}	≤39 mm	>39 mm	0.623	0.591	−0.129	0.702	0.678	+0.006
KO-1		original	$\lambda = 0.2$		0.685	0.861	−0.033	0.635	−16.769	+1.087
	median	all			0.590	−3.286	+0.827	0.567	−21.898	+1.781
		P_{5d}	≤35 mm	>35 mm	0.601	−4.208	+0.855	0.569	−31.133	+1.700
		P_{10d}	≤42 mm	>42 mm	0.532	−7.993	+0.597	0.541	−25.939	+1.597
	mean	all			0.599	0.323	+0.195	0.639	−9.941	+1.408
		P_{5d}	≤35 mm	>35 mm	0.645	−1.217	+0.651	0.59	−25.167	+2.112
		P_{10d}	≤42 mm	>42 mm	0.631	−3.244	+0.469	0.462	−16.854	+1.517
KO-2		original	$\lambda = 0.2$		0.973	−0.133	+1.234	0.845	0.365	+0.501
	median	all			0.656	−0.012	+0.256	0.659	−5.681	+0.835
		P_{5d}	≤41 mm	>41 mm	0.654	−0.093	+0.273	0.621	−11.888	+1.002
		P_{10d}	≤50 mm	>50 mm	0.656	0.083	+0.211	0.638	−10.24	+0.933
	mean	all			0.775	0.758	+0.095	0.716	−3.027	+1.007
		P_{5d}	≤41 mm	>41 mm	0.707	0.410	+0.263	0.716	−3.053	+1.012
		P_{10d}	≤50 mm	>50 mm	0.713	0.502	+0.210	0.817	−2.779	+1.166
KO-3		original	$\lambda = 0.2$		-	−26.299	+0.263	0.888	0.869	+0.057
	median	all			0.661	0.525	+0.170	0.663	−1.663	+0.716
		P_{5d}	≤20 mm	>20 mm	0.649	0.501	+0.180	0.644	−1.779	+0.754
		P_{10d}	≤28 mm	>28 mm	0.649	0.533	+0.161	0.647	−1.509	+0.719
	mean	all			0.676	0.669	+0.057	0.748	−0.273	+0.745
		P_{5d}	≤20 mm	>20 mm	0.673	0.665	+0.056	0.631	−0.719	+0.674
		P_{10d}	≤28 mm	>28 mm	0.673	0.669	+0.040	0.640	−0.251	+0.543

4.2.2. Interpolated λ

Besides discrete values, λ can be expressed as a linear function of P [66,67]. The linear λ–P regressions were calculated for each catchment with and without AMC classification. Analogously to the previous approach (Section 4.2.1), regression equations were calculated for datasets delimited by the P_{5d} and P_{10d} thresholds determined through cluster analysis. All the regression equations are provided in Figure 4 and Table 7.

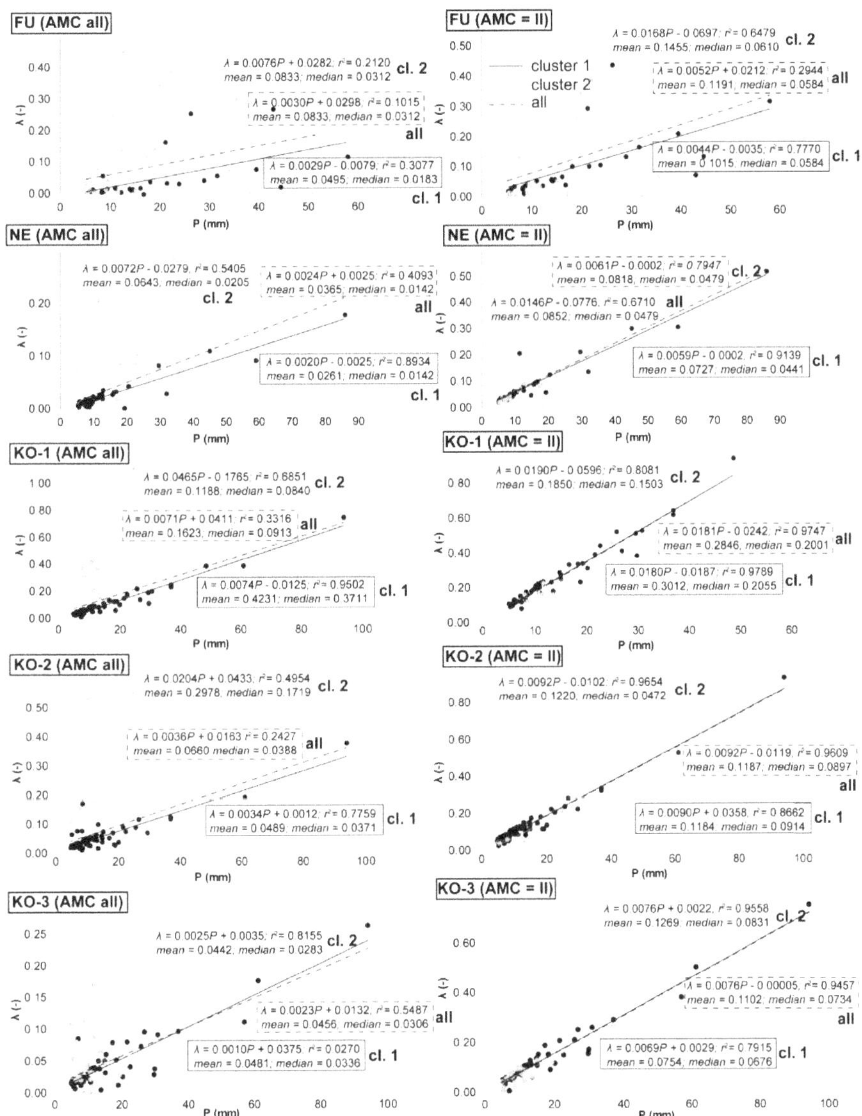

Figure 4. Initial abstraction coefficient λ expressed as linear function of precipitation height P for the experimental catchments and for AMC classified (AMC all) and normal only (AMC = II).

Table 7. Accuracy of direct runoff simulations using regressions of λ according to P in the experimental catchments with (AMC all) and without (AMC = II) determination of AMC classes. $NSE \geq 0.500$ indicating sufficient accuracy are highlighted.

				AMC All			AMC = II		
				r^2	NSE	e	r^2	NSE	e
FU		original	$\lambda = 0.2$	0.258	0.152	−0.818	0.854	0.361	−0.876
	int.	all		0.920	0.670	−0.430	0.633	0.600	+0.035
		P_{5d}	≤26 mm / >26 mm	0.290	0.157	−0.860	0.328	0.312	+0.068
		P_{10d}	≤47 mm / >47 mm	0.290	0.159	−0.870	0.334	0.320	+0.067
NE		original	$\lambda = 0.2$	0.010	−0.070	−0.298	0.781	0.388	−0.253
	int.	all		0.120	−0.740	−0.110	0.688	0.263	−0.197
		P_{5d}	≤18 mm / >18 mm	0.740	0.67	−0.040	0.461	0.357	−0.016
		P_{10d}	≤39 mm / >39 mm	0.580	0.54	−0.030	0.514	0.380	−0.025
KO-1		original	$\lambda = 0.2$	0.010	0.861	+0.005	0.635	−16.789	+1.087
	int.	all		0.140	−0.820	−0.160	0.461	0.347	−0.151
		P_{5d}	≤35 mm / >35 mm	0.390	0.336	−0.120	0.427	0.333	−0.141
		P_{10d}	≤12 mm / >12 mm	0.350	0.313	0.010	0.127	0.339	0.138
KO-2		original	$\lambda = 0.2$	0.970	−0.130	+1.234	0.845	0.365	+0.501
	int.	all		0.070	−0.100	−0.330	0.58	0.322	−0.890
		P_{5d}	≤41 mm / >41 mm	0.340	0.243	−0.120	0.585	0.310	−0.096
		P_{10d}	≤50 mm / >50 mm	0.310	0.229	−0.110	0.42	0.326	−0.074
KO-3		original	$\lambda = 0.2$	-	−26.300	+0.263	0.882	0.876	−0.095
	int.	all		0.500	0.386	−0.200	0.653	0.433	−0.227
		P_{5d}	≤20 mm / >20 mm	0.630	0.536	−0.130	0.566	0.507	−0.109
		P_{10d}	≤28 mm / >28 mm	0.690	0.620	−0.110	0.555	0.452	−0.144

Generally, application of the λ–P linear regressions increased the runoff estimates' accuracy. However, for higher Q the estimates showed systematically underestimating tendency (see Figure 5). Moreover, the effect differed between the experimental catchments. In the FU catchment, the best performance ($NSE = 0.670$) was recorded with using regression for the entire dataset (*all*-i.e., without the separation according to P_{5d} and P_{10d}). In the other four catchments, the cluster-based regressions (i.e., those considering the separation according to P_{5d} and P_{10d}) delivered the best outputs. However, only in catchments NE and KO-3 did NSE values exceed 0.500 (specifically, 0.670 and 0.540 in the NE catchment, 0.536 and 0.620 in the KO-3 catchment). Applying the *all* regression in the NE catchment reduced NSE from −0.070 (original CN method) to −0.740. In the KO-3 catchment, the effect was opposite as NSE increased from −26.3 to 0.386. In the remaining catchments KO-1 and KO-2 the results were similar, but NSE values did not indicate sufficient accuracy. Better results were obtained from simulations that omitted the AMC classification in general (see Figure S2), but NSE exceeded 0.500 only in the FU catchment (*all* regression—$NSE = 0.600$) and the KO-3 catchment (cluster-based P_{5d} regression—$NSE = 0.507$).

Compared with the "discrete λ" approach, the regression-based simulation appears to be more stable in terms of overall enhancement of the CN method's performance. The results indicate that λ should be a variable and event-dependent value. However, the irregular arrangement of the results hints that reliable usage of robust regression-based approach will require a subsequent detailed research of more rainfall-runoff data from various watersheds.

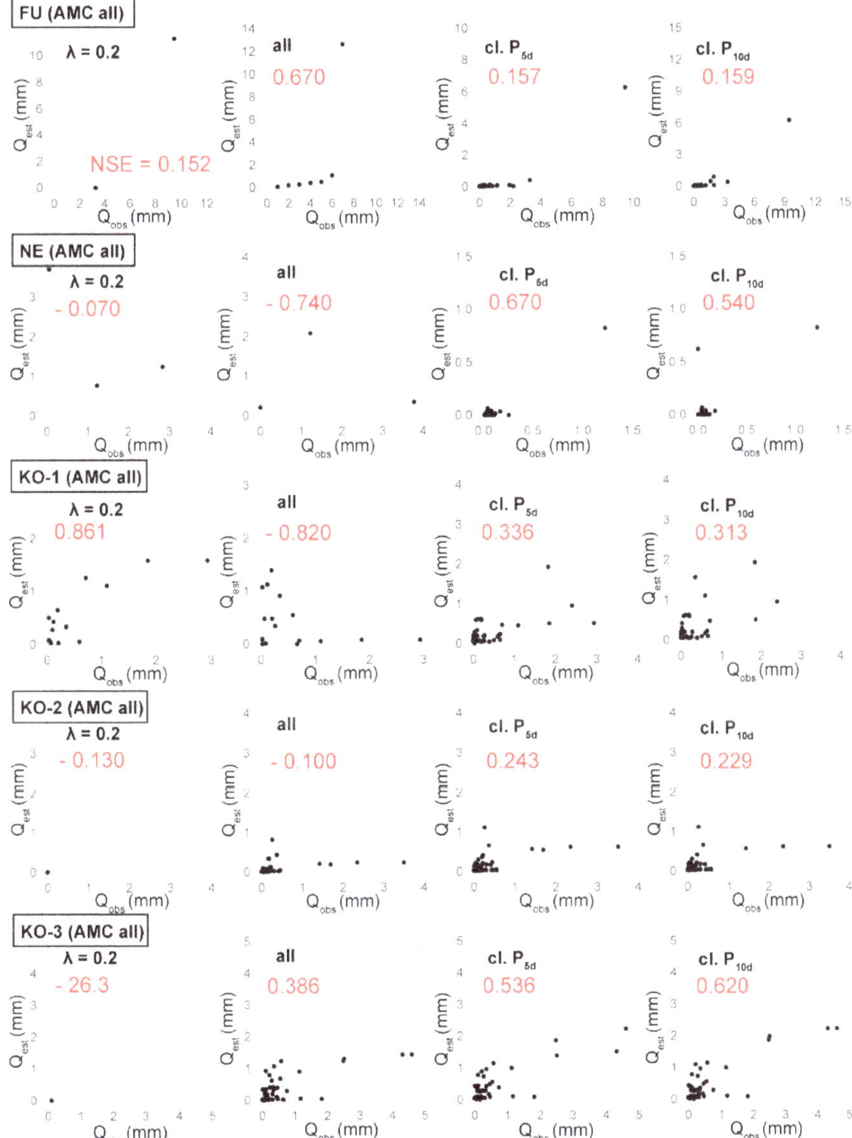

Figure 5. Comparison of observed direct runoff heights (mm) with those simulated in the experimental catchments using regressions of λ according to P for events with AMC classified (AMC all). Comparison of direct runoff estimates with AMC not classified (AMC = II) is provided in Figure S2.

4.3. Approaches Not Dependent on Tabulated CNs

4.3.1. Event Analysis

Unlike the previous analyses, this approach does not start with determining CNs in the watersheds. Instead, it is based on analysing the rainfall-runoff events' parameters (P, Ia, Q), from which λ and S are derived. This procedure aims to analyse the array of λ and S to reveal a natural relation between them.

PCA revealed that P_{10d} poses the most important variable in the arrangement of events (see Figure 3). Therefore, regressions for S according to P_{10d} were calculated for each catchment. Mean and median λ identified in the training dataset were used (see Table 8).

Table 8. Accuracy of direct runoff simulations using regressions of S according to P_{10d} (event analysis) in the experimental catchments using $\lambda = 0.2$, mean and median λ. $NSE \geq 0.500$ indicating sufficient accuracy are highlighted.

Catchment	λ		Regression of S	r^2	NSE	e
FU	orig.	0.2	$S = 13987.982\ P_{10d}^{-1.154}$ $(r^2 = 0.3731)$	-	0.129	−0.820
	mean	0.0183		0.84	0.683	−0.120
	median	0.0029		0.84	0.692	+0.219
NE	orig.	0.2	$S = -318.995\ ln\ (P_{10d}) + 1584{,}638$ $(r^2 = 0.3573)$	0.37	0.371	−0.350
	mean	0.0077		0.75	0.592	+0.206
	median	0.0026		0.74	0.636	−0.100
KO-1	orig.	0.2	$S = 9641.4\ P_{10d}^{-0.859}$ $(r^2 = 0.3473)$	0685	0.861	−0.033
	mean	0.0086		0.6	0.536	−0.040
	median	0.0012		0.67	0.511	+0.070
KO-2	orig.	0.2	$S = 1843.7\ e^{-0.025\ P10d}$ $(r^2 = 0.1750)$	-	−0.133	+1.234
	mean	0.0064		0.91	0.622	−0.170
	median	0.0		0.85	0.758	−0.040
KO-3	orig.	0.2	$S = 874.92\ e^{-0.026\ P10d}$ $(r^2 = 0.2671)$	-	−26.3	+0.263
	mean	0.0129		0.85	0.159	−0.780
	median	0.0010		0.77	0.521	−0.260

Unlike the tabulated *CN*-based approaches, the event analysis delivered more accurate direct runoff estimates (see Table 8, Figure 6). Again, the KO-1 catchment showed lower *NSE* values (0.536 and 0.511) than for the original CN method applied (0.861). On the other hand, the informative value of *NSE* in case of the traditional CN method with $\lambda = 0.2$ was addressed already (Section 4.2.1). Despite the overall improvement of the estimates, there is an underestimating tendency for higher Q evident, similar to the one seen in the tabulated *CN*-based approach.

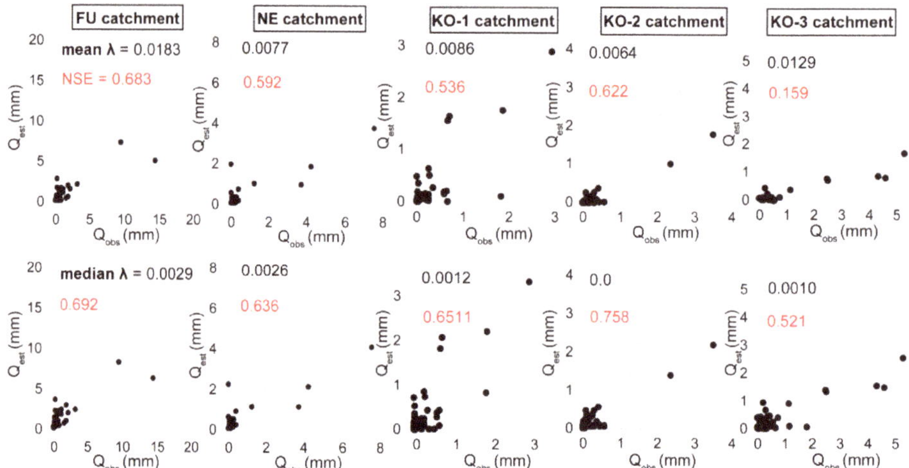

Figure 6. Comparison of observed direct runoff heights with those simulated in the experimental catchments using regressions of S according to P_{10d} (event analysis) with application of mean and median λ.

The results presented in Table 8 show that using P_{10d}-S regressions, along with modified λ, improved direct-runoff estimates' accuracy in all five studied catchments. *NSE* increased (for $\lambda = 0.2$, *NSE* ranged from -2.012 to 0.129) both with mean λ (*NSE* ranged from 0.159 to 0.683) and median λ (*NSE* ranged from 0.511 to 0.758). The observed-vs-modelled runoffs are plotted in Figure 6.

4.3.2. Model Fitting

The iterative model-fitting procedure was applied to training datasets in all studied catchments to identify pairs of S and λ values. As in previous analyses, computations were performed with and without the clustering (separation), according to the P_{5d} and P_{10d} thresholds. The $S{:}\lambda$ pairs are provided in Table 9. Verification has shown that using the identified $S{:}\lambda$ pairs leads to enhancement of estimated direct runoff, compared with those calculated through the original CN method. However, performance differs among catchments (Table 10, Figure 7). For simulations without the separation (*all*), *NSE* ranged from 0.146 to 0.585. *NSE* exceeded 0.500 in the FU and KO-3 catchments. For the P_{5d} -based separation, *NSE* ranged from 0.362 to 0.671, and *NSE* was higher than 0.500 in the FU and KO-3 catchments. Finally, for the P_{10d}-based separation, *NSE* ranged from 0.391 to 0.768, with *NSE* exceeding 0.500 in the NE and KO-3 catchments. In addition, here, the underestimating tendency is apparent for higher Q.

Table 9. Pairs of S and λ values identified by the model fitting iterative procedure in the experimental catchments. Clustered (cl.) and non-clustered (all) events were used. The cluster delimitation was done using the P_{5d} and P_{10d} thresholds.

			S (CN)/λ	
FU	all		371.5 mm (40.64)/0.0	
	cl.	P_{5d}	≤26 mm: 687.5 mm (26.98)/0.0	>26 mm: 138.9 mm (64.65)/0.0
		P_{10d}	≤47 mm: 596.9 mm (29.85)/0.0	>47 mm: 97.9 mm (72.18)/0.0
NE	all		1093.0 mm (18.86)/0.0	
	cl.	P_{5d}	≤18 mm: 1154.5 mm (18.03)/0.0	>18 mm: 12.9 mm (95.17)/1.0
		P_{10d}	≤39 mm: 1154.4 mm (10.03)/0.0	>39 mm: 87.7 mm (74.33)/0.0667
KO-1	all		3949.4 mm (6.04)/0.0	
	cl.	P_{5d}	≤35 mm: 4015.7 mm (5.95)/0.0	>35 mm: 909.9 mm (21.82)/0.0
		P_{10d}	≤42 mm: 4152.3 mm (5.75)/0.0	>42 mm: 1299.2 mm (16.35)/0.0
KO-2	all		2967.0 mm (7.89)/0.0	
	cl.	P_{5d}	≤41 mm: 3037.0 mm (7.72)/0.0	>41 mm: 488.3 mm (34.22)/0.0
		P_{10d}	≤50 mm: 3125.7 mm (7.52)/0.0	>50 mm: 990.0 mm (20.42)/0.0
KO-3	all		905.0 mm (21.92)/0.0	
	cl.	P_{5d}	≤20 mm: 899.8 mm (22.01)/0.0	>20 mm: 495.8 mm (33.88)/0.0
		P_{10d}	≤28 mm: 923.3 mm (21.57)/0.0	>28 mm: 501.0 mm (33.64)/0.0

Table 10. Accuracy of direct runoff simulations using pairs of S and λ identified by means of model fitting procedure. *NSE* ≥ 0.500 indicating sufficient accuracy are highlighted.

	FU Catchment			NE Catchment			KO-1 Catchment			KO-2 Catchment			KO-3 Catchment		
	All	cl.		All	cl.		All	cl.		All	cl.		All	cl.	
		P_{5d}	P_{10d}		P_{5d}	P_{10d}		P_{5d}	P_{10d}		P_{5d}	P_{10d}		P_{5d}	P_{10d}
r^2	0.642	0.56	0.508	0.271	0.799	0.813	0.402	0.494	0.666	0.665	0.521	0.573	0.747	0.796	0.718
NSE	0.585	0.55	0.488	0.247	0.463	0.763	0.173	0.362	0.496	0.146	0.453	0.391	0.547	0.671	0.581
e	−0.062	−0.149	+0.075	−0.195	+0.075	−0.055	−0.253	−0.21	−0.193	−0.149	−0.167	−0.17	+0.642	−0.193	−0.222

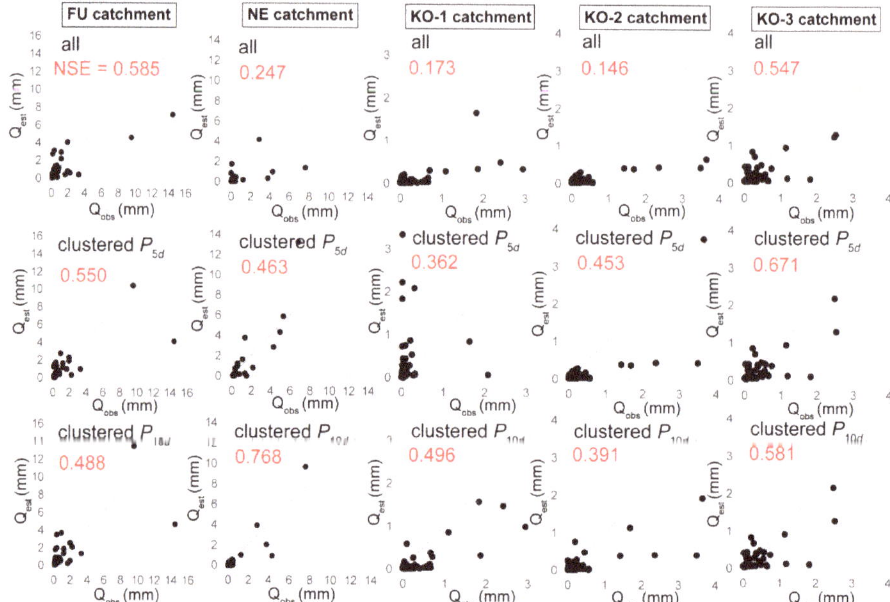

Figure 7. Comparison of observed direct runoff heights with those simulated using pairs of S and λ identified by means of model fitting procedure. The validation was carried out both with (all) and without delimitation of clusters according to P_{5d} (clustered P_{5d}) and P_{10d} (clustered P_{10d}).

4.4. HEC-HMS Simulations

To simulate a surface-runoff process using HEC-HMS, calibration was conducted for two types of events: (1) Those with a slowly rising limb of the hydrograph, and (2) those with a rapid onset. In the calibrations, all parameters' values (see Section 3.5 were optimised, and calibration coefficients were calculated for Ia, CN and T_c (Table 11). The remaining parameters' values were taken as they had been calculated in the calibrations.

Table 11. Calibration coefficients for transformation of Ia, CN and T_c in the three sub-basins of the NE catchment. The coefficients are provided for runoff events with both slow and rapid onset and for the $\lambda = 0.2$ and $\lambda = 0.0142$.

λ	Sub-Basin	Slower Onset			Rapid Onset		
		Ia	CN	T_c	Ia	CN	T_c
	NE-1	0.235	1.000	1.963	0.349	1.010	0.214
0.2	NE-2	0.186	1.000	1.364	0.348	1.010	0.225
	NE-3	0.192	1.000	3.284	0.353	1.325	0.307
	NE-1	0.778	1.015	1.136	0.134	0.670	0.210
0.0142	NE-2	0.778	1.015	0.794	0.136	0.669	0.147
	NE-3	0.750	1.014	1.504	0.179	0.850	0.348

The HEC-HMS simulations indicate that changes in all surface-runoff-driving parameters led to improved results. For the event with the rising limb of the hydrograph rising rather slowly, significant improvement was achieved even for $\lambda = 0.2$ with modified parameters. In the original setting, the bias was −52.29% and NSE was −4.655 (see Figure 8a). In the simulation that employed the modified parameters, bias was only −3.17%, and NSE reached 0.918 (see Figure 8b). For the same event, a simulation using $\lambda = 0.0142$ with the original parameters' values delivered solid output, with bias at -7.01% and *NSE* reaching 0.845 (see Figure 8c). Finally, computations using $\lambda = 0.0142$ and the recalculated parameters led to improvement in the simulation, as bias fell to -2.12%, and *NSE* increased to 0.912 (see Figure 8d).

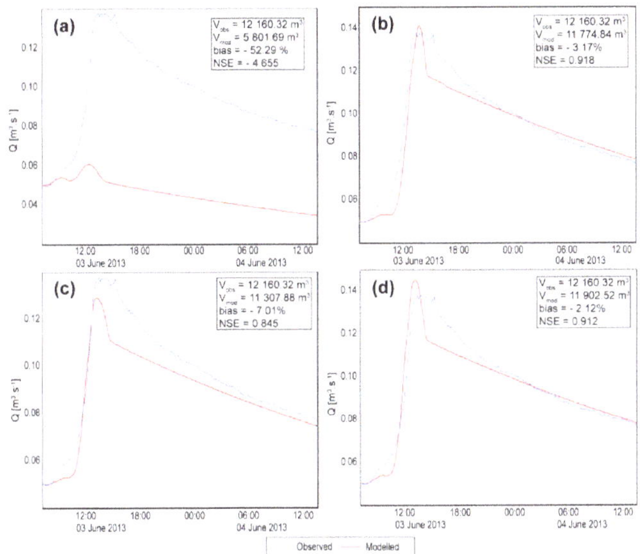

Figure 8. Verification of surface runoff simulations of a runoff event with a slower onset in the NE catchment using the HEC-HMS software. The simulations were computed using original *Ia*, *CN* and T_c values corresponding to $\lambda = 0.2$ (**a**), modified *Ia*, *CN* and T_c values corresponding to $\lambda = 0.2$ (**b**), original *Ia*, *CN* and T_c values corresponding to $\lambda = 0.0142$ (**c**) and finally modified *Ia*, *CN* and T_c values corresponding to $\lambda = 0.0142$ (**d**). The model's performance is expressed by the bias of runoff heights and Nash–Sutcliffe model efficiency (*NSE*) coefficient.

The simulation of runoff response with a rapid onset was less accurate, as indicated in Figure 9. The computation employing $\lambda = 0.2$ with the original setting of the surface-runoff-driving parameters was insufficient (see Figure 9a). After recalculating the parameters, accuracy increased (*NSE* = 0.408), but absolute bias worsened. The modelled runoff depicted in Figure 9b shows an approximation to the real data in terms of temporal course. Analogous use of $\lambda = 0.0142$ delivered a more pronounced improvement in the modelled runoff, with bias falling from 55% to 3%, and *NSE* increasing from -0.194 to 0.695 (Figure 9c,d).

Bearing in mind that the number of simulations by the HEC-HMS performed in this study is low, one should avoid general interpretations. Nevertheless, the presented results outline how reducing *Ia* improves the model's ability to deliver more accurate output, which is significantly more pronounced in the case of slow onset runoff, in which both the temporal course and runoff volume/level matched well with real runoff. In the case of rapid onset events, reducing λ from 0.2 to 0.0142 played a more significant role, as the best simulation was the one with $\lambda = 0.0142$ and recalculated parameters. However, the simulated hydrograph did not match well with the real results (Figure 9d). This indicates that the HEC-HMS simulation of such types of runoff events can deliver sufficiently accurate hydrographs in terms of runoff volume/level, although the culmination is inaccurate.

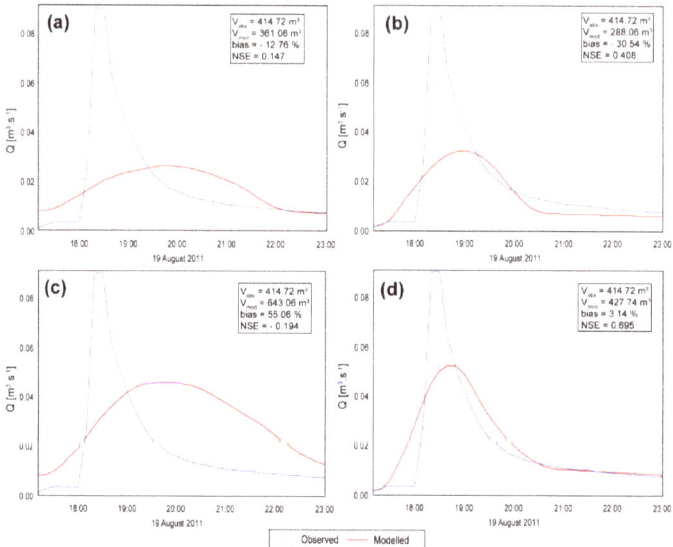

Figure 9. Verification of surface runoff simulations of a runoff event with a rapid onset in the NE catchment using the HEC-HMS software. The simulations were computed using original *Ia*, *CN* and T_c values corresponding to $\lambda = 0.2$ (**a**), modified *Ia*, *CN* and T_c values corresponding to $\lambda = 0.2$ (**b**), original *Ia*, *CN* and T_c values corresponding to $\lambda = 0.0142$ (**c**) and modified *Ia*, *CN* and T_c values corresponding to $\lambda = 0.0142$ (**d**). The model's performance is expressed by the bias of runoff heights and *NSE* coefficient.

5. Discussion and Conclusions

Experience from various watersheds worldwide has highlighted the sensitivity of runoff to the initial abstraction coefficient [2,24,64]. In this study, initial abstraction coefficients λ were estimated for five experimental catchments in the Czech Republic, which well reflect the physiographic conditions (in terms of soil and geomorphology) in the Central European watersheds, using various approaches. From a methodological perspective, they can be divided into two types: (1) Approaches based on a classical determination of tabulated *CN* numbers [65], and (2) those based on a *CN*-independent analysis of rainfall-runoff data. The adjusted λ values were determined as discrete or function-dependent (interpolated). Contribution of this research lies in the comparison of different approaches of *CN* estimations applied in various watersheds.

5.1. Tabulated CN, Discrete λ

When using discrete λ values, these general conclusions based on our findings can be stated: λ differed both between studied catchments and between events, and the original value of 0.2 should be reduced in studied catchments, except the KO-1 catchment. Mean λ ranged from 0.0365 to 0.8330 and median λ from 0.0142 to 0.0913 for simulations that take the AMC classification into consideration. In simulations that omitted the AMC classification (AMC implicitly was set to II), λ values were slightly higher. Mean λ ranged from 0.0818 to 0.2846, and median λ ranged from 0.0479 to 0.2001; λ exceeded 0.2 in the KO-1 catchment, which is almost entirely covered by arable and land with specific infiltration conditions due to presence of tile drainage system [66], making this catchment very different from the others. The example of the KO-1 catchment well documents that correct *CN* determination must take into consideration artificial elements that are not reflected in the traditional methodology, yet their influence on runoff formation is significant.

Generally, the obtained results correspond with those of other authors, such as Baltas et al. [62], who reported mean $\lambda = 0.034$ and 0.014 for two watersheds. They attributed the various λ results to differences in the presence of impervious formations and urban development. Similarly, Shi et al. [11] reported λ ranging from 0.010 to 0.152, with a median of 0.048 due to landscape and geological characteristics. Lal et al. [13] conducted research in different agricultural plots in India and found that the mean and median for λ were 0.0 and 0.034. Krajewski et al. [24] compared λ values in 2 catchments in Poland and found average $\lambda = 0.026$ and 0.047.

5.2. Tabulated CN, Interpolated λ

Some approaches exist for expressing λ as being functionally (linearly) dependent on precipitation. For example, Kohnová et al. [69] conducted research in Slovakian watersheds and expressed λ as a linear function of precipitation for each AMC class. Moreover, they also discussed the low number of II and III AMC classes. Therefore, they stated that no apparent relationship existed between *CN* and AMC, and that other parameters' roles should be emphasised. Application of linear λ:*P* relationships in this study increased runoff estimates' overall accuracy compared with the original CN method. Like the discrete λ, *NSE* only exceeded 0.500 in one catchment—the KO-3 catchment with AMC classified, and in the KO-1 catchment without AMC classification. Generally speaking, the ability of λ:*P* linear regressions used for runoff simulations appears to be rather weak, which can be attributed to unknown interplay among other characteristics taking part in the direct-runoff formation, or to inappropriate tabulated *CN*s. Moreover, a question arises of the need of training datasets large enough, from which representative regression equations can be derived. Despite the sufficient amount of training events in this study (from 44 to 80), there is a problem of lack of extreme/severe events, in which the runoff formation may be very specific.

Recently, a different interpolating approach has been successfully tested by Krajewski et al. [24], approximating the λ–*P* relationship with an asymptotic formula. They found greater variability in λ values for smaller *P*. After exceeding a certain threshold *P*, λ approaches a constant value. The threshold depends on a catchment's characteristics, namely land cover that strongly influences the direct runoff formation or initial abstraction, separately. Based on the promising results of this asymptotic approach, the method definitely deserves to be tested in wider range of geographical conditions.

5.3. Summary of Tabulated CN-Based Approaches

Modification of λ in calculations using tabulated *CN*s led to a partial increase in the accuracy of direct-runoff estimates, but because *NSE* often did not exceed 0.500, overall performance seems to be rather unambiguous. More specifically, it is apparent that the estimates for smaller *Q* were more accurate than for those with higher *Q*. This issue might be resolved based on sophisticated statistical analyses that were beyond the scope of this study.

Both discussed approaches' inaccuracy indicates that the tabulated *CN*s [65] used in the Czech Republic often are not representative, thereby requiring a systematic revision that should be based on analyses of a large amount of data from various watersheds in the Central European region, ensuring the *CN* method's robustness. Moreover, just as it seems inappropriate to use the universal initial abstraction coefficient λ value for different river basins, it certainly would be more appropriate to use the *CN* method, taking into account the spatio-temporal variability of runoff conditions resulting from seasonal variability in *CN* and λ regarding vegetation variability and distribution of soil and land use [24,30–32]. Extant studies have dealt with this issue, such as one conducted by Hawkins et al. [2], who concluded that lower *CN*s prevailed during the summer (growing season) and that higher *CN*s better represented runoff conditions during the dormant season. In a recent study by Muche et al. [79], *CN*s were derived using the Normalised Difference Vegetation Index (NDVI) to reflect seasonal land-use changes in the hydrologic circle. They concluded that the CN-NDVI method is a promising alternative to the traditional CN method. In the present study, the CORINE Land Cover dataset was refined using aerial images together with annual data on crop distribution on arable land, where available. It is reasonable to assume that by using appropriate satellite data, the estimated *CNs* would better represent the spatial distribution of land-cover features, vegetation and also soil moisture. However, this went beyond the scope of this research. Moreover, it should be noted that the Central European region currently is facing the bark beetle calamity and a drought affecting large areas of forest stands (mostly with spruces). This should be taken into account during determination of *CNs*, although the calamity did not affect the studied catchments. In this regard, it should also be added that the issue of a single *CN* value characterising the entire catchment's conditions has been addressed by Soulis and Valiantzas [30,31] who implemented a two-*CN* system approach in heterogeneous watersheds and improved the CN method's performance. Even in a watershed affected by massive changes caused by a forest fire, this approach performed well [32]. Thus, efficient methods exist, offering opportunity to be used in the process of new *CN* determination (revision), not only in the Czech Republic.

5.4. Tabulated CN-Independent Approaches

To estimate optimal λ and S values, two methods without using tabulated *CNs* were applied: (1) Event analysis based on analysing individual events' rainfall-runoff parameters, and (2) model fitting—an iterative numerical procedure searching for a pair of λ and S values that best describe a particular catchment's runoff conditions.

The event analysis yielded the best results from all the methods compared in this research. In all studied catchments, the application of S-P_{10d} regressions combined with mean (ranging from 0.0064 to 0.0129) and median λ (ranging from 0.0 to 0.0029) led to significant increases in estimated direct-runoff accuracy. With one exception (KO-3 catchment with mean λ applied), *NSE* always exceeded 0.500.

The model-fitting procedure also produced more accurate outputs compared with the original CN method, both with and without using the clustering (separation) according to the P_{5d} and P_{10d} thresholds. *NSE* exceeded 0.500 in only two out of the five studied catchments. This probably is caused by the fact that using a single pair of λ and S values that universally represent runoff conditions in a catchment in a wide range of rainfall and land-cover arrangements is questionable [24,30]. Another problem resides in the lack of events with high Q, of which the behaviour is strongly variable, and thus significantly influences the calculation of the λ:S pairs.

5.5. HEC-HMS Simulations Using Adjusted λ and CN

Application of adjusted λ values in HEC-HMS shows that using lower λ leads to improvements in simulated runoff accuracy, but a difference exists between hydrographs with slow and rapid onsets: Both yielded sufficiently accurate results in terms of runoff volume. However, within the hydrograph with rapid onset, the model was unable to match the culmination discharge. This might indicate a lower performance in the HEC-HMS SCS-CN-based simulations to model events with fast runoff response, especially flash floods, in terms of peak discharge, which often is the parameter of the main interest. At the same time, it should be emphasised that general interpretations would require an analysis on a representative dataset of rainfall-runoff events. In this present study, the aim was to examine preliminarily the applicability of modified λ values within HEC-HMS software.

5.6. Summary

The present study shows that the original CN methodology without adjustments to local conditions is hardly applicable in the geographical conditions of the experimental watersheds. Both *CN* and λ values should be based on site-specific characteristics, i.e., on rainfall-runoff data observations, hinting that they should be viewed as flexible regional parameters, rather than fixed values. In the studied catchments' specific natural conditions, the λ appears mostly to be considerably lower than the original value of 0.2. As the CN method frequently is used in long-term and event-based runoff simulations, erosion modelling and wide range of other applications, appropriate modifications of the method, such as those presented in this study, can contribute significantly to improvements in their outputs. It is the authors' intention to extent the present research and take up the unresolved problems, such as the systematic underestimation of direct runoff for events with higher runoff observed. Moreover, testing the possibilities of revisiting *CN* determination, mainly with regard to watersheds' heterogenity, as well as investigation of the regionalised concepts in the natural and agricultural conditions of the Czech Republic and neighbouring countries appears to be a challenging issue deserving our attention. This study's outcomes may serve as a preliminary basis for subsequent activities addressing the aforementioned tasks.

Supplementary Materials: The following are available online at http://www.mdpi.com/2073-4441/12/7/1964/s1, Figure S1a: Comparison of observed direct runoff heights (mm) with those simulated in the experimental catchments using discrete mean and median λ values with AMC classified (AMC all), Figure S1b: Comparison of observed direct runoff heights (mm) with those simulated in the experimental catchments using discrete mean and median λ values with AMC implicitly set as normal (AMC = II), Figure S2: Comparison of observed direct runoff heights (mm) with those simulated in the experimental catchments using regressions of λ according to P for events with AMC classified (AMC all) and implicitly set as normal (AMC = II).

Author Contributions: Conceptualisation, M.C.; methodology, M.C., M.Š.M.; validation, M.C.; formal analysis, M.C., M.Š.M.; P.F.; investigation, M.C.; writing-original draft preparation, M.C., M.Š.M.; writing-review and editing, M.C., M.Š.M., P.F., P.K.; visualisation, M.C. All authors have read and agreed to the published version of the manuscript.

Funding: This research was funded by the DEPARTMENT OF GEOGRAPHY, FACULTY OF SCIENCE, MASARYK UNIVERSITY, grant number MUNI/A/1251/2019, T. G. MASARYK WATER RESEARCH INSTITUTE, internal grant number 1628 and 3600.52.27/2020 and by the MINISTRY OF AGRICULTURE OF THE CZECH REPUBLIC, grant number MZE RO0218.

Acknowledgments: The authors would like to thank the Povodí Odry State Enterprise and Research Institute for Soil and Water Conservation for providing the input rainfall and runoff data. The authors also would like to thank the anonymous reviewers for their constructive comments and recommendations, which helped to improve the paper.

Conflicts of Interest: The authors declare no conflict of interest.

References

1. Bonta, J.V. Determination of Watershed Curve Number Using Derived Distributions. *J. Irrig. Drain. Eng.* **1997**, *123*, 28–36. [CrossRef]
2. Woodward, D.E.; Hawkins, R.H.; Jiang, R.; Hjelmfelt, A.T., Jr.; Van Mullem, J.A.; Quan, Q.D. Runoff curve number method: Examination of the initial abstraction ration. In Proceedings of the World Water and Environmental Resources Congress, Philadelphia, PA, USA, 23–26 June 2003; Bizier, P., De Barry, P., Eds.; pp. 691–700.
3. USDA. *Soil Conservation Service, National Engineering Handbook Supplement: Supplement A, Hydrology*; USDA: Washington, DC, USA, 1956.
4. USDA. *Soil Conservation Service, National Engineering Handbook. Section 4: Hydrology*; USDA: Washington, DC, USA, 1972; p. 762.
5. USDA. *Urban Hydrology for Small Watersheds*, 2nd ed.; Technical Release 55; USDA: Washington, DC, USA, 1986; p. 164.
6. USDA. *National Engineering Handbook, Part 630, Hydrology*; USDA: Washington, DC, USA, 2004.
7. Hawkins, R.H.; Ward, T.J.; Woodwards, E.; Van Mullem, J.A. Continuing evolutioon of rainfall-runoff and the curve number precedent. In Proceedings of the 2nd Federal Inegracy Conference, Las Vegas, NV, USA, 27 June–1 July 2010.
8. Mishra, S.K.; Vijay, P.; Singh, P.K. Revisiting the Soil Conservation Service curve number method. In *Proceedings of the International Conference on Water, Environment, Energy and Society, Bhopal, India, 15–18 March 2016*; Singh, V.P., Yadav, S., Yadava, R.N., Eds.; Springer: Singapore, 2018; pp. 667–693.
9. Ponce, V.M.; Hawkins, R.H. Runoff curve number: Has it reached maturity? *J. Hydrol. Eng.* **1996**, *1*, 11–19. [CrossRef]
10. Satheeshkumar, S.; Venkateswaran, S.; Kannan, R. Rainfall-runoff estimation using SCS-CN and GIS approach in the Pappiredipatti watershed of the Vaniyar sub basin, South India. *Mod. Earth Sys. Environ.* **2017**, *3*. [CrossRef]

11. Shi, Z.H.; Chen, L.D.; Fang, N.F.; Qin, D.F.; Cai, C.F. Research on the SCS-CN initial abstraction ratio using rainfall-runoff event analywsis in the Three Gorges Area, China. *Catena* **2009**, *77*, 1–7. [CrossRef]

12. Jacobd, J.H.; Srinivasan, R. Effects of curve number modification on runoff estimation using WSR-88D raifall data in Texas watersheds. *J. Soil Water Conserv.* **2005**, *60*, 274–279.

13. Lal, M.; Mishra, S.K.; Pandey, A. Physical verification of the effect of land features and antecedent moisture on runoff curve number. *Catena* **2015**, *133*, 318–327. [CrossRef]

14. Springer, E.P.; McGurk, B.J.; Hawkins, R.H.; Coltharp, G.B. Curve numbers from watershed data. In *Proceeding of the Symposium on Watershed Management, Boise, ID, USA, 21–23 July 1980*; ASCE: New York, NY, USA, 1980; pp. 938–950.

15. Mishra, S.K.; Chaudhary, A.; Shrestha, R.K.; Pandey, A.; Lal, M. Experimental Verification of the Effect and Land Use on SCS Runoff Curve Number. *Water Resour. Manag.* **2014**, *28*, 3407–3416.

16. Hawkins, R.H.; Ward, T.J.; Woodwards, E.; Van Mullem, J.A. *Curve Number Hydrology: State of the Practice*; ASCE: Reston, VA, USA, 2009; p. 106.

17. Jha, M.K.; Chovdary, V.M.; Kulkarni, Y.; Mal, B.C. Rainwater harvesting planning using geospatial techniques and multicriteria decision analysis. *Res. Conserv. Recycl.* **2014**, *83*, 96–111. [CrossRef]

18. Sharpley, A.N.; Williams, J.R. EPIC Erosion/Productivity Impact Calculator: 1. Model Documentation. In *USA Department of Agriculture Technical Bulletin No. 1768*; Government Printing Office: Washington, DC, USA, 1990.

19. Ajmal, M.; Moon, G.; Ahn, J.; Kim, T. Investigation of SCS-CN and its inspired modified models for runoff estimation in South Korean watersheds. *J. Hydro-Environ. Res.* **2015**, *9*, 592–603. [CrossRef]

20. Ajmal, M.; Khan, T.A.; Kim, T.W. A CN-Based Ensembled Hydrological Model for Enhanced Watershed Runoff Prediction. *Water* **2016**, *8*, 20. [CrossRef]

21. Mishra, S.K.; Singh, V.P. SCS-CN-based hydrologic simulation package. In *Mathematical Models in Small Watershed Hydrology*; Singh, V.P., Frevert, D.K., Eds.; Water Resources Publications: Littleton, CO, USA, 2002; pp. 391–464.

22. Hjelmfelt, A.T. Ivestigation of curve number procedure. *J. Hydraul. Eng.* **1991**, *117*, 725–737. [CrossRef]

23. Randusová, B.; Marková, R.; Kohnová, S.; Hlavčová, K. Comparison of cn estimation approaches. *Int. J. Eng. Sci.* **2015**, *1*, 34–40.

24. Krajewski, A.; Sikorska-Senoner, A.E.; Hejduk, A.; Hejduk, L. Variability of the Initial Abstraction Ratio in an Urban and an Agroforested Catchment. *Water* **2020**, *12*, 471. [CrossRef]

25. Hjelmfelt, A.T. Empirical investigation of curve number technique. *J. Hydraul. Div. ASCE* **1980**, *106*, 1471–1476.

26. Hawkins, R.H. Discussion of 'Empirical Ivestigation of Curve Number Technique', by A. T. Hjelmfelt. *J. Hydraul. Div. ASCE* **1981**, *107*, 953–954.

27. Hawkins, R.H.; Hjelmfelt, A.T.; Zevenbergen, A.W. Runoff probability, relative storm depth, and runoff curve numbers. *J. Irrig. Drain. Eng.* **1985**, *111*, 330–340. [CrossRef]

28. Hawkins, R.H. 1993 Asymptotic determination of runoff curve numbers from data. *J. Irrig. Dran. Div. ASCE* **1993**, *119*, 334–345. [CrossRef]

29. Ajmal, M.; Kim, T. Quantifying excess stormwater using SCS-CN-based rainfall runoff models and different curve number determination methods. *J. Irrig. Drain. Eng.* **2015**, *141*. [CrossRef]

30. Soulis, K.X.; Valiantzas, J.D. SCS-CN parameter determination using rainfall-runoff data in heterogenous watersheds—The two-CN system approach. *Hydrol. Earth Syst. Sci.* **2012**, *16*, 1001–1015. [CrossRef]

31. Soulis, K.X.; Valiantzas, J.D. Identification of the SCS-CN Parameter Spatial Distribution Using Rainfall-Runoff Data in Heterogenous Watersheds. *Water. Resour. Manage.* **2013**, *27*, 1737–1749. [CrossRef]

32. Soulis, K.X. Estimation of SCS Curve Number variation following forest fires. *Hydrol. Sci. J.* **2018**, *63*, 1332–1346. [CrossRef]

33. Rees, H.W.; Chow, T.L.; Gregorch, E.G. Soil and crop responses to long-term potato production at a benchmark site in northwestern New Brunswick. *Can. J. Soil Sci.* **1988**, *88*, 409–422. [CrossRef]

34. Wang, X.; Liu, T.; Yang, W. Development of a robust runoff-prediction model by fusing the Rational Equation and modified SCS-CN method. *Hydrol. Sci. J.* **2012**, *57*, 1118–1140. [CrossRef]

35. Mishra, S.K.; Pandey, R.P.; Jain, M.K.; Singh, V.P. A rain duration and modified AMC-dependent SCS-CN procedure for long duration rainfall-runoff events. *Water Resour. Manag.* **2008**, *22*, 861–876. [CrossRef]

36. Neitsch, S.L.; Arnold, J.G.; Kiniry, J.R.; Williams, J.R.; King, K.W. *Soil and Water Assessment Tool (SWAT): Theoretical Documentation, Version 2000*; Texas Water Resources Institute: College Station, TX, USA, 2002; p. 506.

37. Sobhani, G. A Review of Selected Small Watershed Design Methods for Possible Adoption to Iranian Conditions. Master's Thesis, Utah State University, Logan, UT, USA, 1975.

38. Mishra, S.K.; Singh, V.P.; Sansalone, J.J.; Aravamuthan, V. A modified SCS-CN method: Characterization and testing. *Water Resour. Manag.* **2003**, *17*, 37–68. [CrossRef]

39. Ponce, V.M. *Engineering Hydrology, Principles and Practices*; Prentice Hall: Englewood Cliffs, NJ, USA, 1989; p. 640.

40. Williams, J.R.; LaSeur, W.V. Water yield model using SCS curve numbers. *J. Hydraul. Div.* **1976**, *102*, 1241–1253.

41. Hawkins, R.H. Runoff curve numbers with varying site moisture. *J. Irrig. Drain. Div.* **1978**, *104*, 389–398.

42. Mishra, S.K.; Jain, M.K.; Singh, V.P. Evaluation of the SCS-CN-based model incorporating antecedent moisture. *Water Resour. Manag.* **2004**, *18*, 567–589. [CrossRef]

43. Mishra, S.K.; Singh, V.P. Long-term hydrological simulation based on the Soil Conservation Service curve number. *Hydrol. Process.* **2004**, *18*, 1291–1313. [CrossRef]

44. Michel, C.; Andréassian, V.; Perrin, C. Soil Conservation Service Curve Number method: How to mend a wrong soil moisture accounting procedure? *Water Resour. Res.* **2005**, *41*. [CrossRef]

45. Jain, M.K.; Durbude, D.G.; Mishra, S.K. Improved CN-based long-term hydrologic simulation model. *J. Hydrol. Eng.* **2012**, *17*, 1204–1220. [CrossRef]

46. Sahu, R.K.; Mishra, S.K.; Eldho, T.I.; Jain, M.K. An advanced soil moisture accounting, procedure for SCS curve number method. *Hydrol. Process.* **2007**, *21*, 2872–2881. [CrossRef]

47. Singh, P.K.; Mishra, S.K.; Berndtsson, R.; Jain, M.K.; Pandey, R.P. Development of modified SMA based MSCS-CN model for Runoff Estimation Water. *Resour. Manag.* **2015**, *29*, 4111–4127.

48. Verma, S.; Mishra, S.K.; Singh, A.; Singh, P.K.; Verma, R.K. An enhanced SMA based SCS-CN inspired model for watershed runoff prediction. *Environ. Earth Sci.* **2017**, *76*. [CrossRef]

49. Verma, S.; Singh, P.K.; Mishra, S.K.; Jain, S.K.; Berndtsson, R.; Singh, A.; Verma, R.K. Simplified SMA-inspired 1-parameter SCS-CN model for runoff estimation. *Arab. J. Geosci.* **2018**, *11*. [CrossRef]

50. Desmukh, D.S.; Chaube, U.C.; Hailu, A.E.; Gudeta, D.A.; Kassa, M.T. Estimation and comparison of curve numbers based on dynamic land use cover change, observed rainfall-runoff data and land slope. *J. Hydrol.* **2013**, *492*, 89–101. [CrossRef]

51. Huang, M.; Gallichand, J.; Wang, Z.; Goulet, M. A modification to the Soil Conservation Service curve number method for steep slopes in the Loess Plateau of China. *Hydrol. Process.* **2006**, *20*, 579–589. [CrossRef]

52. Plummer, A.; Woodward, D.E. The origin and Derivation of Ia/S in the Runoff Curve Number System. In Proceedings of the International Water Resources Engineering Conference, Memphis, TN, USA, 3–7 August 1998; Abt, S.R., Yoing-Pezeshk, J., Watson, C.C., Eds.; ASCE: Reston, VA, USA, 1998; pp. 1260–1265.

53. Fu, S.; Zhang, G.; Wang, N.; Luo, L. Initial abstraction ratio in the SCS-CN method in the Loess Plateau of China. *Trans. ASABE* **2011**, *54*, 163–169. [CrossRef]

54. Fan, F.; Deng, Y.; Hu, X.; Weng, Q. Estimating Composite Curve Number Using an Improved SCS-CN Method with Remotely Sensed Variables in Guangzhou, China. *Remote Sens.* **2013**, *5*, 1425–1438. [CrossRef]

55. Reistetter, J.A.; Russell, M. High-resolution land cover datasets, composite curve numbers, and storm water retention in the Tampa Bay, FL region. *Appl. Geogr.* **2011**, *31*, 740–747. [CrossRef]

56. Mockus, V. Design Hydrographs. In *National Engineering Handbook*; Section 4, Hydrology; McKeever, V., Owen, W., Rallison, R., Eds.; Soil Conservation Service: Washington, DC, USA, 1972.

57. Cazier, D.J.; Hawkins, R.H. Regional application of the curve number method. Water Today and Tomorrow. In Proceedings of the ASCE, Irrigation and Drainage Division Special Conference, Flagstaff, AZ, USA, 24–26 July 1984; Replogle, J.A., Ed.; ASCE: New York, USA, 1984.

58. Mishra, S.K.; Singh, V.P. Another look at SCS-CN method. *J. Hydrol. Eng.* **1999**, *4*, 257–264. [CrossRef]

59. Hawkins, R.H.; Khojeini, A.V. Initial Abstraction and Loss in the Curve Number Method Hydrol; Arizona-Nevada Academy of Science. *Water Resour. Ariz. Southwest* **2000**, *30*, 29–35.

60. Jiang, R. Investigation of Runoff Curve Number Initial Abstraction Ratio. Master's Thesis, The University of Arizona, Tucson, AZ, USA, 2001.

61. Lim, K.J.; Engel, B.A.; Muthukrishnan, S.; Harbor, J. Effects of initial abstraction and urbanization of estimated runoff using CN technology. *J. Am. Water Resour. Assoc.* **2006**, *42*, 629–643. [CrossRef]

62. Baltas, E.A.; Dervos, N.A.; Mimikou, M.A. Technical Note: Determination of the SCS initial abstraction ratio in an experimental watershed in Greece. *Hydrol. Earth Sys. Sci.* **2007**, *11*, 1825–1829. [CrossRef]

63. Xiao, B.; Wang, Q.H.; Fan, J.; Han, F.P.; Dai, Q.H. Application of the SCS-CN model to runoff estimation on a small watershed with high spatial heterogeneity. *Pedosphere* **2011**, *21*, 738–749. [CrossRef]

64. Yuan, Y.; Nie, W.; McCutcheon, S.C.; Taguas, E. Initial abstraction and curve numbers for semiarid watersheds in Southeastern Arizona. *Hydrol. Process.* **2014**, *28*, 774–783. [CrossRef]

65. Janeček, M.; Dostál, T.; Dufková, J.K.; Dumbrovský, M.; Hůla, J.; Kadlec, V.; Konečná, J.; Kovář, P.; Krása, J.; Kubátová, E.; et al. *Ochrana zemědělské půdy před erozí*; Czech University of Life Sciences: Prague, Czech Republic, 2012; p. 113.

66. Caletka, M.; Honek, D. Improving direct runoff estimations through modifying SCS-CN initial abstraction ratio in a catchment prone to flash floods. In Proceedings of the International Multidisciplinary Scientific GeoConference Surveying Geology and Mining Ecology Management, lbena, Bulgaria, 28 June–7 July 2019; pp. 281–288.

67. Caletka, M.; Michalková, M.Š. Determination of SCS-CN initial Abstraction ratio in a chatchment prone to flash floods. *Pollack Per.* **2020**, *15*, 112–123.

68. Karabová, B. Testovanie možnosti regionalizácie vybraných parametrov metódy SCS-CN—Oblasť nížin Slovenska. *Acta Hydrol. Slovaca* **2013**, *15*, 194–203.

69. Kohnová, S.; Karabová, B.; Hlavčová, K. On the possibilities of watershed parametrization of extreme flow estimation on ungauged basins. In Proceedings of the Congress in Flood Risk and Precipitation in Catchments and Cities, Prague, Czech Republic, 22 June–2 July 2015; Chen, Y., Estupina, V.B., Schumann, A., Aksoy, H., Bloschl, G., Kooy, M., Rogger, M., Toth, E., Eds.; Copernicus GmbH: Göttingen, Germany, 2015; pp. 171–175.

70. Kohnová, S.; Rutkowska, A.; Banasik, K.; Hlavčová, K. The L-moment based regional approach to curve numbers for Slovak and Polish Carpathian catchments. *J. Hydrol. Hydromech.* **2020**, *68*, 1–10. [CrossRef]

71. Daňhelka, J.; Kubát, J. *Vyhodnocení Povodní v Červnu a Červenci 2009 (Assessment of Floods in June and July 2009 in the Czech Republic)*; Ministry of Environment, CHMI: Prague, Czech Republic, 2009; p. 165.

72. Šunka, Z. *Vyhodnocení Povodní v Květnu a Červnu 2010 (Assessment of Floods in May and June 2010)*; Ministry of Environment, TGM WRI: Prague, Czech Republic, 2010; p. 172.

73. Honek, D.; Michalková, M.Š.; Smetanová, A.; Sočuvka, V.; Velísková, Y.; Karásek, P.; Konečná, J.; Némethová, Z.; Danáčová, M. Estimating sedimentation rates in small reservoirs—Suitable approaches for local municipalities in central Europe. *J. Environ. Manag.* **2020**, *261*. [CrossRef] [PubMed]

74. Goovaerts, P. Geostatistical approaches for incorporating elevation into the spatial interpolation of rainfall. *J. Hydrol.* **2000**, *228*, 113–129. [CrossRef]

75. Yang, K.; Cao, S.; Liu, X. Flow resistance and its precipitation methods in compound channels. *Acta Mech. Sin.* **2007**, *23*, 23–31. [CrossRef]

76. Pham, D.T.; Dimov, S.S.; Nguyen, C.D. Selection of K in K-means clustering. *J. Mech. Eng. Sci.* **2005**, *219*, 103–119. [CrossRef]

77. Dingman, S.L. *Physical Hydrology*, 2nd ed.; Macmillan Press Limited: New York, NY, USA, 2002; p. 646.

78. Konečná, J.; Karásek, P.; Beitlerová, H.; Fučík, P.; Kapička, J.; Podhrázská, J.; Kvítek, T. Using WaTEM/SEDEM and HEC-HMS episodic models for the simulation of episodic hydrological and erosion events in a small agricultural catchment. *Soil Water Res.* **2019**, *15*, 18–29. [CrossRef]

79. Muche, M.E.; Hutchinson, S.L.; Hutchinson, J.M.S.; Johnston, J.M. Phenology-adjusted dynamic curve number for improved hydrologic modelling. *J. Environ. Manag.* **2019**, *235*, 403–413. [CrossRef]

Article

Possibility of Using Selected Rainfall-Runoff Models for Determining the Design Hydrograph in Mountainous Catchments: A Case Study in Poland

Dariusz Młyński [1,*], Andrzej Wałęga [1], Leszek Książek [2], Jacek Florek [2] and Andrea Petroselli [3]

[1] Department of Sanitary Engineering and Water Management, University of Agriculture in Krakow, St. Mickiewicza 24–28, 30-059 Krakow, Poland; andrzej.walega@urk.edu.pl

[2] Department of Hydraulics Engineering and Geotechnics, University of Agriculture in Krakow, St. Mickiewicza 24–28, 30-059 Krakow, Poland; leszek.ksiazek@urk.edu.pl (L.K.); rmflorek@cyf-kr.edu.pl (J.F.)

[3] Department of Economics, Engineering, Society and Business Organization (DEIM), Tuscia University, Via San Camillo De Lellis snc, 01100 Viterbo, Italy; petro@unitus.it

* Correspondence: dariusz.mlynski@urk.edu.pl; Tel.: (+48)-12-662-4041

Received: 25 April 2020; Accepted: 19 May 2020; Published: 20 May 2020

Abstract: The aim of the study was to analyze the possibility of using selected rainfall-runoff models to determine the design hydrograph and the related peak flow in a mountainous catchment. The basis for the study was the observed series of hydrometeorological data for the Grajcarek catchment area (Poland) for the years 1981–2014. The analysis was carried out in the following stages: verification of hydrometeorological data; determination of the design rainfall; and determination of runoff hydrographs with the following rainfall-runoff models: Snyder, NRCS-UH, and EBA4SUB. The conducted research allowed the conclusion that the EBA4SUB model may be an alternative to other models in determining the design hydrograph in ungauged mountainous catchments. This is evidenced by the lower values of relative errors in the estimation of peak flows with an assumed frequency for the EBA4SUB model, as compared to Snyder and NRCS-UH.

Keywords: design hydrograph; SCS-CN; EBA4SUB; mountainous catchments; rainfall-runoff models; ungauged catchments

---

1. Introduction

Surface runoff is the amount of water that is generated when excess stormwater, meltwater, or other water sources flow over the topographic surface. It occurs, for instance, when the soil is saturated from above by infiltration, or when the soil is saturated from below by the subsurface flow. Surface runoff often occurs due to impervious areas (such as roofs and others) that do not allow water to soak into the ground. It is the primary agent of soil erosion by water [1,2]. In addition to causing soil erosion, surface runoff is a primary cause of flooding, which can result in property damages. Hence, determining the shape of design hydrographs is a very important activity in flood protection.

The knowledge of the characteristics of design hydrograph, such as peak flow, duration, or volume, is necessary for planning and designing water management facilities and determining flood risk zones. One of the methods for obtaining such information is the use of theoretical design hydrographs. These are typical hydrographs for the investigated catchment, which describe the shape of the flood wave. Hydrograph parameters are often determined on the basis of physiographic characteristics of the catchment. One particular application related to water management during floods is in uncontrolled catchments. Indeed this application is challenging because ungauged catchments' lack of runoff measurements that are often necessary for calibrating advanced hydrological models [3,4].

There are quite a number of methods that can be used to determine design hydrographs in ungauged basins, for instance hydrological rainfall-runoff models can be used [5–8]. Of the numerous mathematical models used for the analysis of the rainfall-runoff relationship, conceptual models based on the cascade of Nash linear tanks, double cascade of tanks (Wackermann model) or synthetic unit hydrographs (Snyder, SCS-UH, Clark-UH) are commonly employed [9]. It should be emphasized that in recent years, the runoff formation in a changing environment has become an important scientific problem in hydrology. A better understanding of the runoff changes are thus of paramount importance to effectively manage water resources. The variability of hydro-meteorological conditions due to climate change significantly affects the hydrological regime in catchments [10]. Generally, climate change and human activities are significant factors influencing runoff variation [11,12]. Hence, studies related to the possibility of using new methods to determine design hydrographs should be conducted.

Past experience with the use of rainfall-runoff models has shown some limitations associated with their use. Many problems focus on the following issues: (1) Lack of guidelines for the adoption of a standard rainfall hyetograph, which increases the uncertainty of the results obtained; (2) High sensitivity of synthetic hydrographs to the distribution of rainfall patterns in the catchment and rainfall height measurement errors; (3) Overestimation or underestimation of direct runoff from the original SCS-CN method; (4) Uncertainty of the results of SCS-UH in relation to the input parameters; (5) Subjectivity of the selection of the size of the indicators for estimating the parameters of selected models [13–16]. An alternative to the rainfall-runoff models used so far may be the recently developed "Event-based Approach for Small and Ungauged Basins" (EBA4SUB) model. It allows estimating the magnitude of the peak flow along with the characteristics of the design hydrograph (e.g., duration and volume). This model has been fully adapted to determine the floods in uncontrolled catchments. It is based on geographic information systems and on the optimization of topographic information contained in the digital elevation model. The EBA4SUB model was developed to obtain the design hydrograph with the smallest possible input information like when using the well-known rational formula [17–19].

Considering the limitations related to the previously aforementioned rainfall-runoff models, the overall objective of this study was a comparative analysis for the shape of design hydrograph determined in a Carpathian mountainous catchment with the following models: NRCS-UH, Snyder, and EBA4SUB. The novelties in the EBA4SUB model are in its approach towards calculating excess rainfall and its adaptation of the width function to the description of the transformation of rainfall into runoff. It should be emphasized that so far there has been no research regarding the possibility of using the EBA4SUB model to determine the design hydrographs in the Carpathian catchments. Therefore, a novelty in this study is the comparison of the EBA4SUB model with simple models commonly used in Poland, which have certain recognized limitations. The study will allow a determination as to whether the EBA4SUB model may be an alternative to commonly used methodological approaches.

2. Materials and Methods

The basis for this study was the observed time series of daily rainfall and maximum annual flows in the period 1971–2014 for the Grajcarek river, obtained from the Institute of Meteorology and Water Management in Warsaw, the National Research Institute. The analyses were carried out in the following stages: verification of hydrometeorological data, calculation of maximum annual rainfall and flows with a specified probability, determination of the design hydrographs using the analysed rainfall-runoff models described in the following, and quality assessment of the used models.

2.1. Research Area Characteristic

The selected case study is the Grajcarek river catchments (coordinates for cross-section Szczawnica: 49°25′22″ N 20°28′58″ E). It is located in the Carpathian part of the upper Vistula river catchment, in Poland. The catchment area is 86 km^2. The length of the main watercourse to the outlet cross-section is 15.0 km. The average catchment slope is 8%. The density of the river network is 0.70 km^{-1}.

This information was derived using a Digital Elevation Model (DEM) provided by USGS—United States Geological Survey, with a grid resolution of about 25 m. Based on the Hydrographic Division Map of Poland 2010 and on the Corine Land Cover 2018 database, the following forms of land uses were identified together with the percentage of the catchment area respectively occupied. Loose urban buildings (4.5%), arable lands without irrigation (5.1%), meadows and pastures (14.4%), areas occupied mainly by agriculture with a high proportion of natural vegetation (2.1%), coniferous forests (29.8%), mixed forests (43.4%), and deciduous forests (0.2%). The case study is dominated by poorly permeable and impermeable soils. The average annual rainfall in the catchment area exceeds 765 mm. The average annual temperature is 6.8 °C. Figure 1 shows the land use and the topography of the catchment.

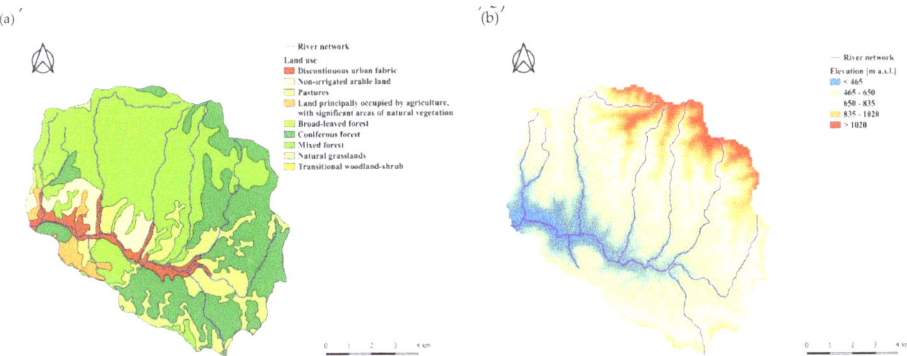

Figure 1. Case study land use (**a**) and topography (**b**).

2.2. Hydrometeorological Data Verification

The verification of the hydrometeorological data was carried out in relation to the maximum annual daily rainfall (P_{max}) and the maximum annual flow (Q_{max}), using the Mann-Kendall test (MK). The H_0 null hypothesis of the test assumes no monotonic data trend, while the H_1 alternative states that such a trend exists. The calculations were done for the significance level $\alpha = 0.05$. The S Mann-Kendall statistics were determined based on the equation [20,21]:

$$S = \sum_{k=1}^{n-1}\sum_{j=k+1}^{n} \text{sgn}(x_j - x_k) \tag{1}$$

$$sgn\,(x_j - x_k) = \begin{cases} 1 \text{ for}(x_j - x_k) > 0 \\ 0 \text{ for}(x_j - x_k) = 0 \\ -1 \text{ for}(x_j - x_k) < 0 \end{cases} \tag{2}$$

where:

n—number of elements in the time series.

The Z normalized statistics were calculated from the equation:

$$Z = \frac{S - \text{sgn}(S)}{Var\,(S)^{1/2}} \tag{3}$$

where: $Var(S)$—variance S is determined from the equation:

$$Var(S) = \frac{1}{18}\cdot(n\cdot(n - 1)\cdot(2n + 5)) \tag{4}$$

The main premise the MK test used was the lack of autocorrelation in the data series. In the case of P_{max} and Q_{max} analysis, such relationships may occur, which in consequence leads to an underestimation of the *Var(S)* variance. Therefore, a correction for correction of variance has been included, calculated only for data with significant partial autocorrelation:

$$Var*(S) - Var(S) \cdot \frac{n}{n_s^*} \tag{5}$$

where:

Var(S)*—corrected variance;
n—the real number of observation;
n_s^*—effective number of observations calculated as:

$$\frac{n}{n_s^*} = 1 + \frac{2}{n(n-1)(n-2)} \cdot \Sigma_{k=1}^{n-1} (n-k)(n-k-1)(n-k-2)\rho_k \tag{6}$$

where:

k—next group with repeating elements;
ρk—value of the next significant autocorrelation coefficient.

2.3. Calculation of Maximum Annual Rainfall and Flows with Specific Occurrence Frequency

In this study, it was assumed that the maximum flow with a specific frequency (Q_T) is caused by the occurrence of rainfall with the same frequency [22]. The Q_T flows were determined in order to assess the quality of the hydrological rainfall-runoff models. The tests were performed for the frequencies related to the return periods of 500, 100, and 10 years. The calculations were performed applying the log-normal distribution, which is described as [23]:

$$f(x) = \frac{1}{(x-\varepsilon)\alpha\sqrt{2\pi}} exp\left[-\frac{1}{2}\left(\frac{ln(x-\varepsilon)-\mu}{\alpha}\right)^2\right] \tag{7}$$

$$x_p = exp\left[\mu + \frac{\alpha\sqrt{2}}{erf(2(1-p)-1)}\right] + \varepsilon \tag{8}$$

where:

x_p—quantile of the theoretical log-normal distribution;
ε—lower string limit;
$erf(2(1-p)-1)$—Gauss error function.

After determining the representative rainfall, the concentration time for the catchment was determined using the Giandotti formula [24]. Rainfall was then determined for a duration equal to the concentration time, using Lambor reduction curves. Then the pattern of the design hyetograph was determined with the DVWK method [25].

2.4. Determination of the Design Hydrograph

The design hydrograph was determined with the models: Snyder, SCS-UH, and EBA4SUB. In the Snyder and SCS-UH models, the excess rainfall was determined using the SCS-CN method, while in EBA4SUB the CN4GA procedure was used. In order to characterize the design hydrograph shape, the values of wave slenderness coefficients were determined from the relationship [26]:

$$\alpha = \frac{t_o}{t_s} \tag{9}$$

where:

t_0—wave fall time (h);
t_s—wave rise time (h);

The SCS-CN method is a common method used for estimating the direct runoff in the world. It was developed in USA, mostly for assessing the runoff in small agricultural catchments [27,28]. In the SCS-CN method, the excess rainfall depends on the category of soils, on the land use, and on the soil moisture before the considered rainfall occurs. All these factors were considered in the *CN* parameter (Curve Number) which is an empirical parameter used in hydrology for predicting direct runoff or infiltration from total rainfall. The ranges of *CN* is from 0 to 100. The amount of excess rainfall was calculated from the relationship [29–31]:

$$Pe = \begin{cases} \frac{(P - 0.2S)^2}{P + 0.8S}, & \text{when } P \geq 0.2S \\ 0, & \text{when } P < 0.2S \end{cases} \tag{10}$$

where:

P_e—excess rainfall (mm);
P—total rainfall (mm);
S—maximum potential catchment retention (mm).

The maximum potential retention of the catchment is directly related to the *CN* parameter and is described by the equation:

$$S = 25.4 \cdot \left(\frac{1000}{CN} - 10 \right) \tag{11}$$

The *CN* parameter was determined for the second moisture level (AMC II), calculating it for the catchment as a weighted average, according to the guidelines presented in [32].

To determine the design hydrograph with the Snyder model, it is necessary to know the peak flow, delay time, and the time to reach the peak. These parameters are estimated from the relationship [33–35]:

$$T_L = C_t \cdot (L \cdot L_c)^{0.3} \tag{12}$$

where:

T_L—delay time (h);
C_t—factor related to catchment retention (-);
L—maximum distance along the watercourse from the outlet cross-section to the drainage divide (km);
L_c—distance along the main watercourse from the outlet cross-section to the centroid of the catchment (km).

$$Q_p = \frac{2.78 \cdot C_p \cdot A}{T_L} \tag{13}$$

where:

Q_p—peak flow of the unit hydrograph ($\text{m}^3 \cdot \text{s}^{-1} \cdot \text{mm}$);
C_p—empirical coefficient resulting from the simplification of the hydrograph to triangular shape (-);
A—catchment area (km^2).

The SCS-UH model belongs to the group of unit hydrograph methods. The unit hydrograph is simplified to form a triangle, and it is caused by an excess unitary rainfall uniformly distributed throughout the entire catchment area. The size of the peak flow is expressed by the formula [36]:

$$q_p = \frac{c \cdot A}{T_p} \tag{14}$$

where:

q_p—peak flow of the unit hydrograph (m^3·s^{-1}·mm);
c—conversion factor ($c = 0.208$) (-);
T_p—flood rise time, (h), calculated as:

$$T_P = \frac{D}{2} + T_{LAG} \tag{15}$$

where:

D—duration of excess rainfall (h);
T_{LAG}—lag time in the SCS-UH method, (h), calculated as:

$$T_{LAG} = \frac{(3.28 \cdot L \cdot 1000)^{0.8} \cdot \left(\frac{1000}{CN} - 9\right)^{0.7}}{1900 \cdot \sqrt{I}} \tag{16}$$

where:

L—maximum length of the runoff path (km);
CN—Curve Number value (-);
I—average catchment slope (%).

The main element determining the shape of the design hydrograph determined with the NRCS-UH method is the Peak Rate Factor (PRF). The value recommended by NRCS is 484. In the case of small mountain catchments characterized by a fast response to rainfall, the value of this indicator should be higher. In this work, analyses using the SCS-UH model were carried out for PRF values equal to 484 and 600.

In order to use the EBA4SUB model to determine the runoff hydrograph, it was necessary to use the DEM of the catchment and to determine the value of the CN parameter (the same of other methodologies). The excess rainfall was subsequently determined based on the CN4GA procedure [37]. This method is based on two stages. In the first, the excess rainfall is determined based on the NRCS-CN method (Equation (10)). In the second stage, the temporal distribution of the excess rainfall inside the event is determined using the Green–Ampt equation:

$$q_0(t) = \begin{cases} i(t), & \text{for } t < t_p \\ K_s\left(1 + \frac{\Delta\theta\Delta H}{I(t)}\right), & \text{for } t > t_p \end{cases} \tag{17}$$

where:

q_0—infiltration indicator;
t_p—ponding time;
K_s—saturated hydraulic conductivity;
I—cumulative infiltration;
$\Delta\theta$—change in soil-water content between the initial value and the field saturated soil-water content;
ΔH—difference between the pressure head at the soil surface and the matric pressure head at the moving wetting front.

The solution of Equation (17) requires the calibration of the K_s parameter. At the beginning, the value of this parameter is assumed based on the case study soil group. Then the cumulative infiltration is calculated, and its value is compared to that obtained from the NRCS-CN method. The value of the K_s parameter changes until the cumulative infiltration from Equation (17) is equal to that calculated using the NRCS-CN method. After determining the amount of excess rainfall, the instantaneous unit hydrograph is determined based on the width function described by the relationship [38]:

$$WFIUH(t) = \frac{L_c(x)}{V_c(x)} + \frac{L_h(x)}{V_h(x)} \tag{18}$$

where:

L_c, L_h—hillslope and channel flow paths, functions of DEM cell x, respectively;

V_c, V_h—runoff velocity for hillslope cells and flow channel cells.

After defining *WFIUH*, the final design hydrograph $Q(t)$ is determined, described by the following relationship:

$$Q(t) = A \int_0^t WFIUH(t - \tau)P_n(\tau)d\tau \tag{19}$$

where:

A—catchment area (km^2);

T—duration of rainfall (h);

$P_n(t)$—excess rainfall determined by the CN4GA method (mm/h).

2.5. Assessment of Quality of Analysed Hydrological Models

The quality of the simulations obtained using the Snyder, NRCS-UH and EBA4SUB models was assessed using the Mean Absolute Percentage Error (MAPE), which is described by the following relationship [39]:

$$MAPE = \frac{Q_{s,max} - Q_{m,max}}{Q_{s,max}} \cdot 100 \ [\%] \tag{20}$$

where:

$Q_{m,max}$—maximum flow with a certain frequency of occurrence, calculated using rainfall-runoff models (m$^3\cdot$s^{-1}),

$Q_{s,max}$—maximum flow with a specified frequency of occurrence, calculated using the log-normal distribution on observed data (m$^3\cdot$s^{-1}).

3. Results

3.1. Hydrometeorological Data Verification

The results of the analysis related to the hydrometeorological data verification are presented in Table 1.

Table 1. The results of the significance analysis of the trend for the observation series P_{max} and Q_{max} in the Grajcarek catchment.

Characteristic	Z_c	p_c	Var_c	n/n *	Z	p	Var
P_{max}	1.582	0.114	7721.554	0.846	1.455	0.146	9129.333
Q_{max}	1.143	0.253	9766.667	1.000	1.143	0.253	9766.667

Z_c—modified value of normalized MK statistics; p_c—modified value of test probability, Var_c—modified value of variance, n/n *—effective number of observations, Z—value of normalized statistics MK, p—test probability, Var—variance.

Based on the results summarized in Table 1, it was found that for the analyzed multi-year period, there were no statistically significant trends in the series of observations of the maximum annual daily rainfall and flow. This is evidenced by the values of the Z_c Mann-Kendall statistics, which assume smaller quantities than the critical value, for the significance level $\alpha = 0.05$ at 1.96. In the case of the results for the P_{max} observation series, the effective number of observations (n/n *) takes values other than 0, which indicates autocorrelation between individual variables. However, it does not affect the conclusions made using the analyzed test. Similar results for rainfall in southern Poland were obtained by Niedźwiedz et al. [40] and Młyński et al. [41]. These studies showed that the rainfall in this region is characterized by growing but not statistically significant trends, and all changes in the amount of rainfall are caused by irregular fluctuations. Slight trends in changes in the amount of rainfall are reflected in the flood flows in the upper Vistula catchment. Research conducted by Kundzewicz et al. [42], Wałęga et al. [43] and Młyński et al. [44] have shown a steady state of flood flows in southern Poland over the last several years. Due to their geological structure, morphometric characteristics, and land use, mountainous basins of the upper Vistula river catchment are sensitive to the occurrence of intense rainfall. Hence, the rhythm of their course is repeated by the rhythm of rainfall [45]. Therefore, it can be pointed out that there is a relationship between the maximum annual daily rainfall and the maximum flow.

3.2. Determination of Rainfall and Peak Flows at a Specific Occurrence Frequency

Rainfall and maximum flow of a certain frequency were determined using the log-normal distribution. The results of the calculations are presented in Figures 2 and 3. Based on the analyses, the following rainfall values have been determined for return periods of 500, 100, and 10 years: 139; 111; and 73 mm respectively. The following values were obtained for peak flows: 157.660; 101.469; and 44.470 $m^3 \cdot s^{-1}$ for return periods of 500, 100, and 10 years, respectively. The use of log-normal distribution in the analysis for calculating the maximum daily annual rainfall with a specific frequency of occurrence was supported by research conducted by Młyński et al. [46] It has been shown that this function is best suited to the empirical rainfall distribution P_{max}, hence it can be the basis for determining the course of design rainfall. In the work Młyński et al. [47] it has been shown also that the same function describes at best the empirical distributions of Q_{max} flows in the upper Vistula catchment. The log-normal distribution belongs to the family of right-handed heavy-tail functions, hence it is widely used to describe extreme hydrometeorological phenomena, as supported by Kuczera [48] and Strupczewski et al. [49]

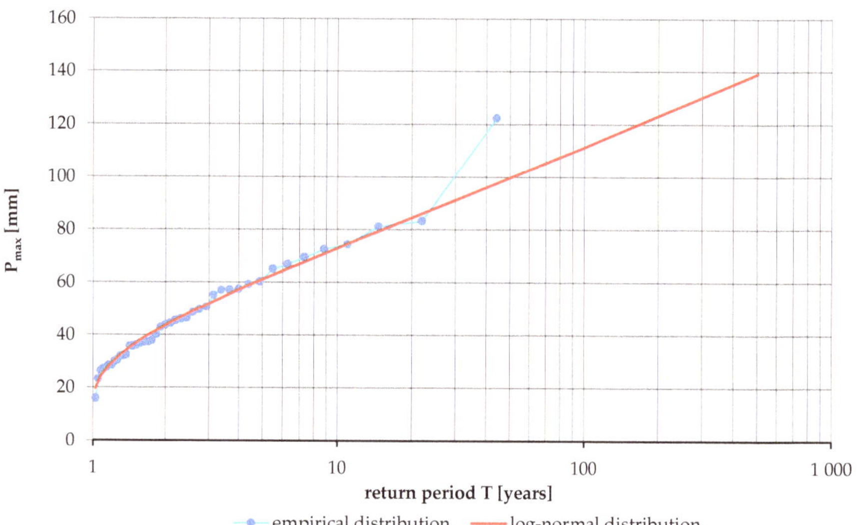

Figure 2. Probability distribution curve for P_{max}.

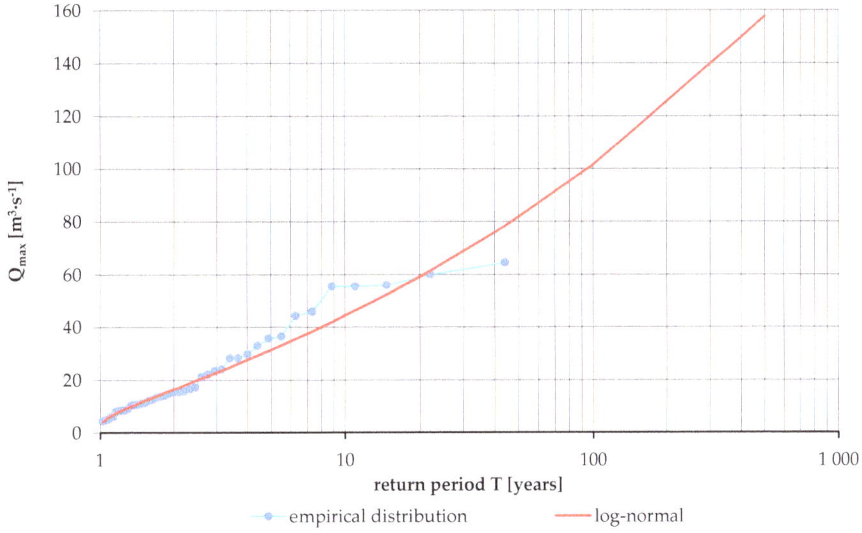

Figure 3. Probability distribution curve for Q_{max}.

3.3. Determination of Design Hydrographs Employing the Selected Rainfall-Runoff Models

In order to be able to use the selected rainfall-runoff models, the excess rainfall heights were determined using the NRCS-CN and CN4GA methods. The results of the calculations are summarized in Table 2. Figure 4 shows the excess rainfall hyetographs against the background of total rainfall.

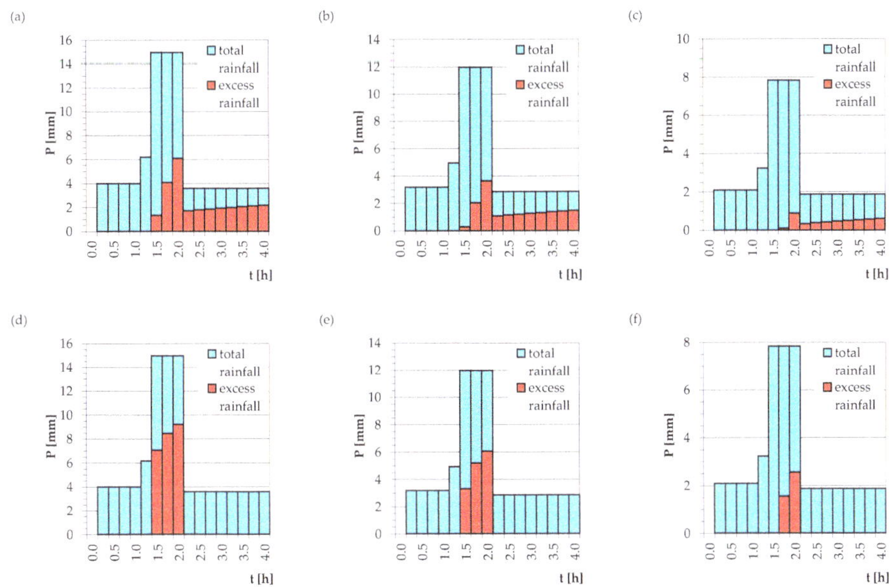

Figure 4. Excess rainfall hyetographs against the background of the total rainfall determined by the following methods: (**a**) NRCS-CN for return period 500 years; (**b**) NRCS-CN for return period 100 years; (**c**) NRCS-CN for return period 10 years; (**d**) CN4GA for return period 500 years; (**e**) CN4GA for return period 100 years; (**f**) CN4GA for return period 10 years.

Table 2. Excess rainfall amounts determined by the NRCS-CN and CN4GA methods.

Return Period	T_c (h)	CN	P (mm)	P_{net} (mm)
500			95.8	24.7
100	4	68.1	76.6	14.6
10			50.2	4.1

The amount of total rainfall P for the assumed return periods was determined by reducing the amounts calculated using the log-normal distribution to an amount corresponding to a duration of 4 h, due to the calculated concentration time. AMC II was assumed for the calculation of the *CN* parameter. It should be emphasized, however, that assuming this level for ungauged mountainous catchments is a significant problem because the initial soil moisture conditions are determined not only by atmospheric rainfall but also by the high level of the groundwater table and the occurrence of poorly permeable soils that hinder infiltration. These factors lead to a decrease in the retention capacity of the catchment. This is particularly evident in mountainous areas. Therefore, the adoption of a low level of moisture can lead to an underestimation of the size of the design hydrograph. If the soil moisture level is set too high, the floods can be greatly overestimated. Therefore, it is necessary to verify the assumptions in relation to local environmental conditions to reduce modeling uncertainty [50]. Moreover, it should be emphasized that using the *CN* values proceeded by the standardized procedure, the SCS-CN method can overestimate runoffs for high rainfall events, whereas it underestimates runoffs for low depth rainfall events [51].

The shape of design hydrograph for given return periods was determined using the Snyder, NRCS-UH, and EBA4SUB models. In the case of the first two models, the analysis was carried out for excess rainfall calculated by the NRCS-UH method. For the EBA4SUB model, the excess rainfall determined by the CN4GA method was used. Table 3 summarizes the characteristics of design hydrographs. Figures 5–7 present their shapes for the investigated return periods.

Table 3. Characteristics of design hydrographs determined using the analyzed models.

Characteristic	Snyder			NRCS-UH			EBA4SUB		
	500	100	10	500	100	10	500	100	10
Q_{max} [m³·s⁻¹]	122.909	74.191	21.975	113.235 * 135.483 **	68.424 * 82.059 **	20.329 * 24.529 **	212.531	125.602	35.779
V [mln m³]	2.291	1.377	0.403	2.291	1.377	0.403	2.078	1.226	0.346
t [h]	15.250	15.250	15.000	21.750 * 16.000 **	21.750 * 15.750 **	21.500 * 15.500 **	6.800	6.800	6.500
α [–]	2.100	1.952	1.905	3.150 * 2.095 **	3.000 * 2.095 **	2.950 * 2.150 **	1.545	1.545	1.500

* PRF (peak rate factor) = 484, ** PRF = 600.

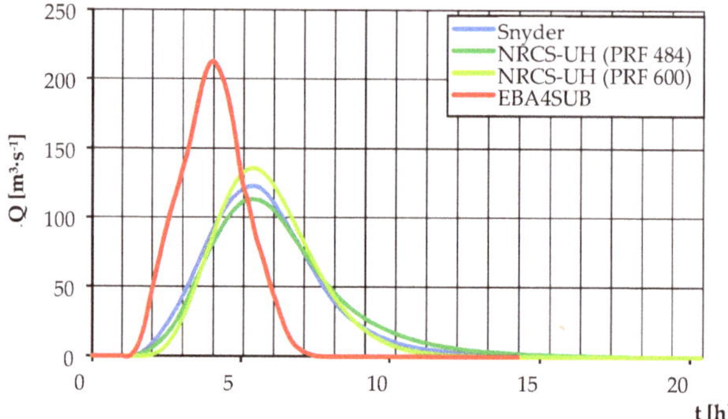

Figure 5. Design hydrograph for *T* = 500 years return period.

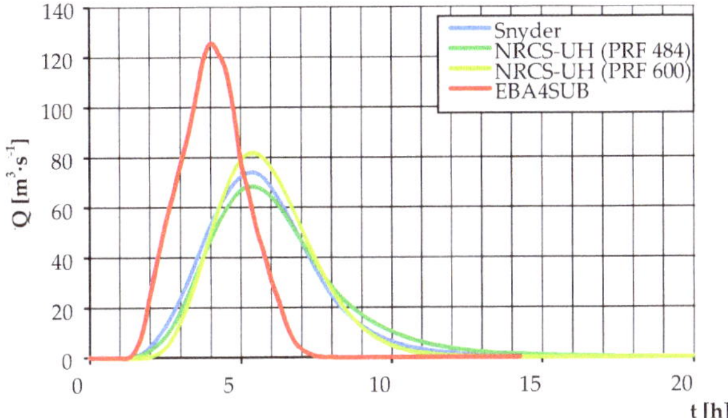

Figure 6. Design hydrograph for T = 100 years return period.

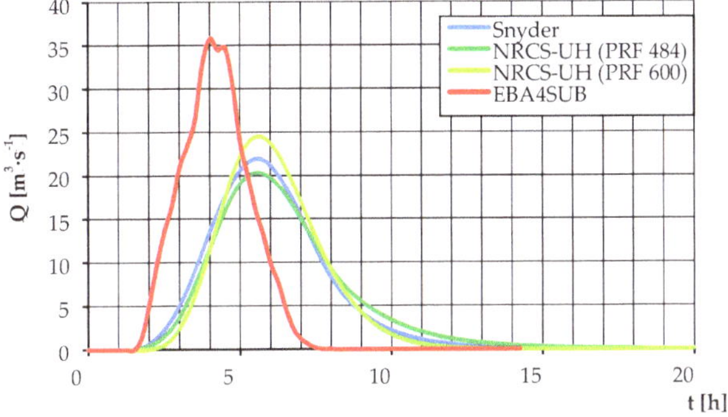

Figure 7. Design hydrograph for T = 10 years return period.

Based on the results summarized in Table 3 and Figures 5–7, it was found that the highest values of peak flows were obtained using the EBA4SUB model. This is due to the fact that the EBA4SUB model takes into account the surface runoff speed when calculating the width function from the formula (18). The analysed catchment has a mountainous character, so the runoff speeds will be higher due to the greater slopes. Therefore, the catchment response for the EBA4SUB model will be faster than for the others. In the NRCS-UH model, the catchment topography is considered indirectly by the PRF value, however, in this case, a generalization is assumed that the same PRF value is valid for the entire catchment, which is typical for a lumped model. In the case of the EBA4SUB model, the surface flow velocity is calculated for each pixel of the DEM, and is dependent on the specific pixel slope, so it characterizes the dynamics of runoff for the entire catchment. For the Snyder model, the topography of the terrain is included in the Ct parameter. However, the model is not very sensitive to changes in this parameter [26]. Regarding the peak flow values for Snyder and NRCS-UH, they are comparable. It should be emphasized that higher values in the peak flow for the NRCS-UH model were obtained assuming the PRF value as 600. Due to their geological structure and physiographic characteristics, small mountainous catchments are characterized by a rapid response to the occurring rainfall. As a result, the peak flow is violent and the flood duration is relatively short. Higher PRF values reflect the shape of such floods. Conversely, lower values should be assumed for lowland and plain catchments [52]. The average peak flows obtained with the Snyder and NRCS-UH models are 40% smaller compared to the corresponding peak flows determined by the EBA4SUB model, for each return period. Despite significant differences in flows between EBA4SUB, Snyder and NRCS-UH, the volume of flood is similar and amounts to slightly over 2 million m^3, 1 million m^3 and ca. 0.4 million m^3 for return periods of 500, 100, and 10 years, respectively. Similar volumes result from the fact that the basic information on excess rainfall in the CN4GA model are values determined using

NRCS-CN. The CN4GA method first calculates the cumulative excess rainfall using the NRCS-CN method. The CN4GA model is then used to determine the distribution of the excess rainfall over time by taking into account the variable infiltration capacity during soil in the catchment. The shorter peak flow time is the result of the distribution of excess rainfall over time, determined by the CN4GA method (Figure 5). The duration of this fallout is 0.75 h (500 and 100 years return period) and 0.5 h (10 years return period) where for the NRCS-CN method it is 2.8 h and 2.5 h, respectively. Therefore, the effective rainfall intensity for CN4GA is concentrated over a shorter period of time, which is reflected by the shape of the design hydrograph. The slenderness factor, whose values in each case are greater than 1.0, indicates that the water level increased rapidly until the peak and then it fell slowly. Hence, it can be concluded that there is an imbalance between volumes for the rising and falling parts, where the falling volume was in each case larger. Due to its simplicity, the EBA4SUB model can be alternatively used to determine the shape of design hydrographs in uncontrolled catchments. Its use consists basically of three stages: determining the distribution of the design rainfall, determining the excess rainfall, and determining the design hydrograph from the catchment. The simplicity of application of the model in hydrological analyses means that it can be successfully used by practitioners to determine flood hazard zones or to determine the size of reliable flows for the design of hydrotechnical constructions. All analysed models have some limitations. They are sensitive to changes in the *CN* parameter, which determines the amount of excess rainfall, which translates into the values of the peak flow. Research carried out by Maidment and Hoogerwerf [53] showed that an increase in the *CN* parameter by 1% increases the peak flow determined from the model also by approximately 1%. In the case of the Snyder model, the coefficient values depending on the retention capacity of the catchment must be estimated for its identification. Currently, there are no specific guidelines indicating which values of these parameters should be assumed in relation to particular characteristics of the catchment, which is why designers do it in a subjective way. The first attempts to make the parameters of the Snyder model dependent on the catchment characteristics in Polish conditions were carried out by Wałęga [54], but the obtained results require verification in the Carpathian catchments. Changes in the values of individual parameters may, as a consequence, affect the sizes of floodplains or planned water management facilities. The conducted research allowed to state that the EBA4SUB model does not have such limitations. Model parameters are determined directly on the basis of physiographic characteristics of the catchment, which precludes the subjectivity of their determination.

However, it should be emphasized that further research should be carried out regarding the possibility of the use of EBA4SUB in uncontrolled mountainous catchments. Such studies should primarily focus on determining the reference shape of the design hyetograph. A research conducted by Młyński et al. [44] concluded, for instance, that even small changes in the design hyetograph shape can cause differences in peak flows, calculated using the model, of more than 10%.

The Grajcarek drainage catchment is strongly asymmetrical, i.e., most of its tributaries are right-bank inflows and are what causes that the shape of the catchment to be close to the oval. Hence the concentration time is relatively short and it can contribute to a relatively fast rise of hydrographs. It is defined very well by the EBA4SUB model compared to classic models, which take less account of catchment asymmetry in the hydrographs shape.

3.4. Evaluation of the Quality of Analysed Hydrological Models

The results of the calculations regarding the quality of hydrological models using the MAPE indicator are presented in Table 4. Figure 8 presents the values of peak flows for the analysed return periods obtained using the analysed rainfall-runoff models against the background of the probability distribution curve determined by the log-normal distribution.

Table 4. Mean Absolute Percentage Error (MAPE) index values for the hydrological models analyzed.

Model	MAPE [%]		
	500	100	10
Snyder	22.0	26.9	50.5
NRCS-UH (PRF 484)	28.2	32.6	54.2
NRCS-UH (PRF 600)	14.1	19.1	44.7
EBA4SUB	−34.8	−23.8	19.4

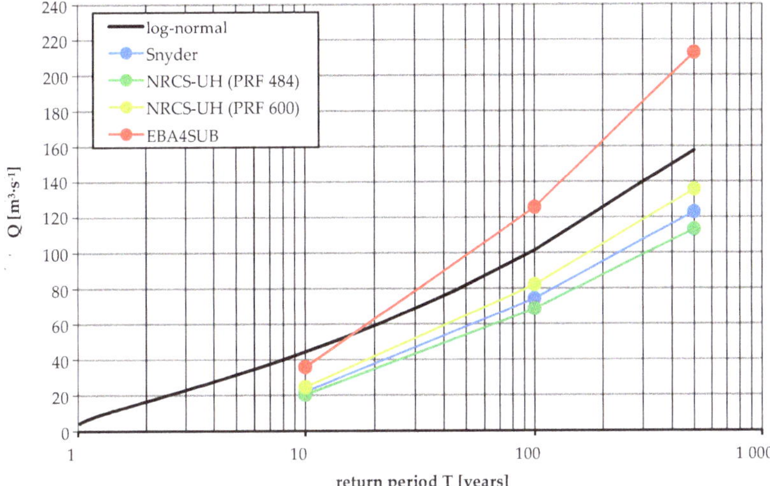

Figure 8. Q_T values obtained with the analysed hydrological models against the background of log-normal distribution.

Based on the results shown in Figure 8, it was found that Q_T values determined using the Snyder and NRCS-UH models are lower, as respect to the corresponding values determined using the log-normal distribution. For the EBA4SUB model, the calculated Q_{500} and Q_{100} are higher than the statistical method. Based on the values listed in Table 4, it was found that for Q_{500}, the NRCS-UH (PRF 600) model was characterized by the smallest relative error. It should be emphasized, however, that the average value of this error, for all return periods, is lower for the EBA4SUB and NRCS-UH (PRF 600) models and amounted to 26% for both. For Snyder it was 33%, while for NRCS-UH (PRF 484) it was 38%. The calculations carried out confirm that the EBA4SUB model can be an alternative to the Snyder and NRCS-UH models. In the case of lower return periods, the EBA4SUB model gives peak flows more similar to the observed ones. This provides some security for uncontrolled catchments, where there is no hydrometric information. This could reduce the risk of designating too-narrow flood hazard zones or undersizing hydrotechnical constructions in the light of weather phenomena, the course of which is becoming increasingly extreme.

The main aim of EBA4SUB is to provide an accurate estimation of the design hydrograph, minimizing at the same time the subjectivity of the practitioner related to the choice of the input parameters. Practically, the EBA4SUB model reduces the uncertainty in the modeling of both the infiltration process (thanks to the automatic estimation of *CN*) and both the propagation process (thanks to the automatic estimation of concentration time). In doing so, EBA4SUB proposes a framework that, on the same watershed and with the same input data, provides similar results when it is applied by two different analysts in two different moments. This is crucial because when using different approaches, as respect to EBA4SUB, like the well-known rational formula, it recognized a great uncertainty in the estimation of the runoff coefficient or the concentration time. Moreover, from a methodological point of view, EBA4SUB is characterized by the following advantages. First, in excess rainfall estimation, thanks to the CN4GA structure and its automatic calibration, it benefits from the accuracy of a physically based infiltration scheme (the Green–Ampt equation) mixed with the simplicity of an empirical approach (the *CN* method). Second, for excess rainfall-direct runoff transformation, it determines the IUH (instantaneous unit hydrograph) shape using detailed geomorphological information pixel by pixel, avoiding the use of synthetic methods.

The proposed procedure can help in assessing the influence of land abandonment on a watershed scale on the surface runoff, since it allows an immediate estimation of design hydrograph based on a hypothesis of changes of *CN* and concentration time that could be due to land use or land cover modifications. Indeed modifications in land use or land cover are reflected by changes in *CN*, affecting the infiltration process, and by changes in concentration time, altering surface flow velocities that at a cascade affect the excess rainfall-direct runoff transformation [55]. Regarding agricultural practices, it is well known in the literature that the impact of agriculture leads to a reduction in vegetation cover and to an increase of both soil erosion rates and flood formation phenomena. As an example, Cerdà et al. [56] found that when a field is abandoned, a sudden increase in surface runoff (more than two times) can be expected before vegetation recovers, so there is a particular need to apply nature-based soil and water conservation strategies to prevent soil erosion. Conversely, regarding forestry practices, Keesstra [57] found that the basin extensive reforestation of the Dragonja catchment, which occurred in Slovenia from 1945 to 2008, reduced the discharge of the river across the entire spectrum of the flow, in particular for high return period events. Finally, regarding urbanization processes, Recanatesi et al. [58] found that uncontrolled land degradation can severely increase flood risk, and that best management practices are strongly needed, since they can help in mitigating the flood risk by up to 90%.

4. Conclusions

The purpose of the research was to analyze the possibility of using the EBA4SUB model to determine design hydrographs in mountainous catchments. Considering the obtained results, it was found that this model can be an alternative to commonly used rainfall-runoff models such as Snyder and NRCS-UH. The EBA4SUB model provides values for peak flows, with a specific frequency of occurrence, relatively similar in respect to values obtained by the statistical method. Therefore, it can be recommended as one of the alternatives to determine the shape of design hydrographs in uncontrolled and mountainous basins. It should be emphasized, however, that the decisive factor influencing the performances of hydrological rainfall-runoff models is the quality of meteorological data constituting the input signal. Therefore, it is recommended to use the model in catchments where this information has been fully verified.

Considering the topographic characteristics of the catchments that affect the asymmetry of its river network, like height differences, runoff length and therefore the time of wave transition, the EBA4SUB model appears to show less subjectivity in runoff calculation processes. Practical verification will be possible by using the model to delineate flood risk zones and compare them with the corresponding results obtained with other methods or with observed ones.

Author Contributions: Conceptualization, D.M., A.W. and A.P.; methodology, D.M. and A.W.; software, A.P.; validation, D.M. and A.W.; formal analysis, D.M. and A.W.; investigation, D.M. and A.W.; resources, D.M. and A.P.; data curation D.M.; writing—original draft preparation, D.M.; writing—review and editing, D.M., A.W., L.K., J.F. and A.P.; visualization, D.M., A.W., X.X., D.M., X.X., D.M., A.W., L.K., J.F. and A.P. All authors have read and agreed to the published version of the manuscript.

Funding: This research was funded by Ministry of Science and High Education in Poland.

Conflicts of Interest: The authors declare no conflict of interest.

References

1. Cerdà Bolinches, A.; Lucas Borja, M.E.; Úbeda, X.; Martínez Murillo, J.F.; Keesstra, S. Pinus halepensis M. versus Quercus ilex subsp. Rotundifolia L. runoff and soil erosion at pedon scale under natural rainfall in Eastern Spain three decades after a forest fire. *For. Ecol. Manag.* **2018**, *400*, 447–456. [CrossRef]
2. Van Eck, C.M.; Nunes, J.P.; Vieira, D.C.; Keesstra, S.; Keizer, J.J. Physically-Based modelling of the post-fire runoff response of a forest catchment in central Portugal: Using field versus remote sensing based estimates of vegetation recovery. *Land Degrad. Dev.* **2016**, *27*, 1535–1544. [CrossRef]
3. Banach, W. Determination of synthetic flood hydrograph in ungauged catchments. *Infrastruct. Ecol. Rural Areas* **2011**, *12*, 147–156.
4. Salami, A.W.; Bilewu, S.O.; Ibitoye, A.B.; Ayanshola, A.M. Runoff hydrographs using Snyder and SCS unit hydrograph methods: A case study of selected rivers in south west Nigeria. *J. Ecol. Eng.* **2017**, *18*, 25–34. [CrossRef]
5. Gądek, W. Assessment of limingraph data usefulness for determining the hypothetical flood waves with the Cracow method. *J. Water Land Dev.* **2014**, *21*, 71–78. [CrossRef]
6. Gądek, W.; Środula, A. The evaluation of the design flood hydrographs determined with the Hydroproject method in the gauged catchments. *Infrastruct. Ecol. Rural Areas* **2014**, *4*, 29–47.
7. Pietrusiewicz, I.; Cupak, A.; Wałęga, A.; Michalec, B. The use of NRCS synthetic unit hydrograph and Wackerman conceptual model in the simulation of a flood wave in uncontrolled catchment. *J. Water Land Dev.* **2014**, *23*, 53–59. [CrossRef]
8. Wypych, A.; Ustrnul, Z.; Henek, E. Meteorological hazards—Visualization system of national protection against extreme hazards for Poland. *Meteorol. Hydrol. Water Manag.* **2014**, *2*, 37–42. [CrossRef]
9. Karabová, B.; Sikorska, A.; Banasik, K.; Kohnová, S. Parameters determination of a conceptual rainfall-runoff model for a small catchment in Carpathians. *Ann. Wars. Univ. Life Sci. SGGW* **2012**, *44*, 155–162.
10. Wałęga, A.; Młyński, D. Seasonality of median monthly discharge in selected Carpathian rivers of the upper Vistula basin. *Carpath. J. Earth Environ. Sci.* **2017**, *12*, 617–628.
11. Tolentino, P.L.M.; Poortinga, A.; Kanamaru, H.; Keesstra, S.; Maroulis, J.; David, C.P.C.; Ritsema, C.J. Projected impact of climate change on hydrological regimes in the Philippines. *PLoS ONE* **2016**, *11*, e0163941. [CrossRef]
12. Xu, J.; Wang, S.; Bai, X.; Shu, D.; Tian, Y. Runoff response to climate change and human activities in a typical karst watershed, SW China. *PLoS ONE* **2018**, *13*, e0193073. [CrossRef] [PubMed]
13. Blair, A.; Sanger, D.; White, D.; Holland, A.F.; Vandiver, L.; Bowker, C.; White, S. Quantifying and simulating stormwater runoff in watersheds. *Hydrol. Process.* **2012**, *28*, 559–569. [CrossRef]
14. Banasik, K.; Rutkowska, A.; Kohnová, S. Retention and Curve Number variability in a small agricultural catchment: The probabilistic approach. *Water* **2014**, *6*, 1118–1133. [CrossRef]
15. Kowalik, T.; Wałęga, A. Estimation of CN Parameter for small agricultural watersheds using asymptotic functions. *Water* **2015**, *7*, 939–955. [CrossRef]
16. Wałega, A.; Książek, L. Influence of rainfall data on the uncertainty of flood simulation. *Soil Water Res.* **2016**, *11*, 277–284. [CrossRef]
17. Grimaldi, S.; Petroselli, A. Do we still need the Rational Formula? An alternative empirical procedure for peak discharge estimation in small and ungauged basins. *Hydrol. Sci. J.* **2015**, *60*, 67–77. [CrossRef]
18. Piscopia, R.; Petroselli, A.; Grimaldi, S. A software package for the prediction of design flood hydrograph in small and ungauged basins. *J. Agric. Eng.* **2015**, *432*, 74–84.
19. Petroselli, A.; Grimaldi, S. Design hydrograph estimation in small and fully ungauged basin: A preliminary assessment of the EBA4SUB framework. *J. Flood Risk Manag.* **2018**, *11*, 197–2010. [CrossRef]
20. Sun, S.; Barraud, S.; Branger, F.; Braud, I.; Castebrunet, H. Urban hydrology trend analysis based on rainfall and runoff data analysis and conceptual model calibration. *Hydrol. Process.* **2017**, *31*, 1349–1359. [CrossRef]
21. Operacz, A.; Wałęga, A.; Cupak, A.; Tomaszewska, B. The comparison of environmental flow assessment—The barrier for investment in Poland or river protection? *J. Clean. Prod.* **2018**, *193*, 575–592. [CrossRef]
22. Viglione, A.; Blöschl, G. On the role of storm duration in the mapping of rainfall to flood return period. *Hydrol. Earth Syst. Sci. Discuss.* **2008**, *5*, 3419–3447. [CrossRef]
23. Hogg, R.V.; Craig, A.T. *Introduction to Mathematical Statistics*; Macmilan Publishing Co.: New York, NY, USA, 1978.

24. Grimaldi, S.; Petroselli, A.; Tauro, F.; Porfiri, M. Time of concentration: A paradox in modern hydrology. *Hydrol. Sci. J.* **2012**, *57*, 217–228. [CrossRef]

25. Wałęga, A.; Drożdżal, E.; Piórecki, M.; Radoń, R. Some problems of hydrology modelling of outflow from ungauged catchments with aspect of flood maps design. *Acta Sci. Pol. Form. Circumiectus* **2012**, *11*, 57–68. (In Polish)

26. Wałęga, A. The importance of calibration parameters on the accuracy of the floods description in the Snyder's model. *J. Water Land Dev.* **2016**, *28*, 19–25. [CrossRef]

27. Soulis, K.X.; Dercas, N. Development of a GIS-based Spatially Distributed Continuous Hydrological Model and its First Application. *Water Int.* **2007**, *32*, 177–192. [CrossRef]

28. Soulis, K.X.; Valiantzas, J.D. Identification of the SCS-CN Parameter Spatial Distribution Using Rainfall-Runoff Data in Heterogeneous Watersheds. *Water Res. Man.* **2013**, *27*, 1737–1749. [CrossRef]

29. Xiao, B.; Wang, Q.; Fan, J.; Han, F.; Dai, Q. Application of the SCS-CN Model to Runoff Estimation in a Small Watershed with High Spatial Heterogeneity. *Pedosphere* **2011**, *26*, 738–749. [CrossRef]

30. Soulis, K.X.; Valiantzas, J.D. SCS-CN parameter determination using rainfall-runoff data in heterogeneous watersheds—The two-CN system approach. *Hydrol. Earth Syst. Sci.* **2012**, *16*, 1001–1015. [CrossRef]

31. Soulis, K.X. Estimation of SCS Curve Number variation following forest fires. *Hydrol. Sci. J.* **2018**, *63*, 1332–1346. [CrossRef]

32. Wałęga, A.; Salata, T. Influence of land cover data sources on estimation of direct runoff according to SCS-CN and modified SME methods. *Catena* **2019**, *172*, 232–242. [CrossRef]

33. Singh, P.K.; Mishra, S.K.; Jain, M.K. A review of the synthetic unit hydrograph: From the empirical UH to advanced geomorphological methods. *Hydrol. Sci. J.* **2014**, *59*, 239–261. [CrossRef]

34. Sudhakar, B.S.; Anupam, K.S.; Akshay, A.J. Snyder unit hydrograph and GIS for estimation of flood for un-gauged catchments in lower Tapi basin, India. *Hydrol. Curr. Res.* **2015**, *6*, 1–10.

35. Federova, D.; Kovář, P.; Gregar, J.; Jelínková, A.; Novotná, J. The use of Snyder synthetic hydrograph for simulation of overland flow in small ungauged and gauged catchments. *Soil Water Res.* **2018**, *13*, 185–192.

36. Amatya, D.M.; Cupak, A.; Wałęga, A. Influence of time concentration on variation of runoff from a small urbanized watershed. *GLL* **2015**, *2*, 7–19.

37. Grimaldi, S.; Petroselli, A.; Romano, N. Curve-Number/Green–Ampt mixed procedure for streamflow predictions in ungauged basins: Parameter sensitivity analysis. *Hydrol. Process.* **2013**, *27*, 1265–1275. [CrossRef]

38. Grimaldi, S.; Petroselli, A.; Nardi, F.; Alonso, G. Flow time estimation with variable hillslope velocity in ungauged basins. *Adv. Water Resour.* **2010**, *33*, 1216–1223. [CrossRef]

39. Kim, S.; Kim, H. A new metric of absolute percentage error for intermittent demand forecast. *Int. J. Forecast* **2016**, *32*, 669–679. [CrossRef]

40. Niedźwiedź, T.; Łupikasza, E.; Pińskwar, I.; Kundzewicz, Z.W.; Stoffel, M.; Małarzewski, Ł. Climatological background of floods at the northern foothills of the Tatra Mountains. *Theor. Appl. Climatol.* **2014**, *119*, 273–284. [CrossRef]

41. Młyński, D.; Cebulska, M.; Wałęga, A. Trends, variability, and seasonality of Maximum annual daily precipitation in the upper Vistula basin, Poland. *Atmosphere* **2018**, *9*, 313. [CrossRef]

42. Kundzewicz, Z.W.; Stoffel, M.; Kaczka, R.J.; Wyżga, B.; Niedźwiedź, T.; Pińskwar, I.; Ruiz-Villanueva, V.; Łupikasza, E.; Czajka, B.; Ballesteros-Canovas, J.A. Floods at the Northern Foothills of the Tatra Mountains—A Polish–Swiss Research Project. *Acta Geophys.* **2014**, *62*, 620–641. [CrossRef]

43. Wałęga, A.; Młyński, D.; Bogdał, A.; Kowalik, T. Analysis of the course and frequency of high water stages in selected catchments of the Upper Vistula basin in the south of Poland. *Water* **2016**, *8*, 394. [CrossRef]

44. Młyński, D.; Petroselli, A.; Wałęga, A. Flood frequency analysis by an event-based rainfall-runoff model in selected catchments of southern Poland. *Soil Water Res.* **2018**, *13*, 170–176.

45. Kundzewicz, Z.W.; Pińskwar, I.; Choryński, A.; Wyżga, B. Floods still pose a hazard. *Aura* **2017**, *3*, 3–9. (In Polish)

46. Młyński, D.; Wałęga, A.; Stachura, T.; Kaczor, G. A new empirical approach to calculating flood frequency in ungauged catchments: A case study of the upper Vistula basin, Poland. *Water* **2019**, *11*, 601. [CrossRef]

47. Młyński, D.; Wałęga, A.; Petroselli, A.; Tauro, F.; Cebulska, M. Estimating maximum daily precipitation in the upper Vistula basin, Poland. *Atmosphere* **2019**, *10*, 43. [CrossRef]

48. Kuczera, G. Robust flood frequency models. *Water Resour. Res.* **1982**, *18*, 315–324. [CrossRef]

49. Strupczewski, W.G.; Singh, V.P.; Mitosek, H.T. Non-stationary approach to at site flood frequency modeling. III. Flood analysis for Polish rivers. *J. Hydrol.* **2001**, *248*, 152–167. [CrossRef]

50. Wałęga, A.; Rutkowska, A. Usefulness of the modified NRCS-CN method for the assessment of direct runoff in a mountain catchment. *Acta Geophys.* **2015**, *63*, 1423–1446. [CrossRef]

51. Soulis, K.X.; Valiantzas, J.D.; Dercas, N.; Londra, P.A. Investigation of the direct runoff generation mechanism for the analysis of the SCS-CN method applicability to a partial area experimental watershed. *Hydrol. Earth Syst. Sci.* **2009**, *13*, 605–615. [CrossRef]

52. Walega, A.; Amatya, D.M.; Caldwell, P.; Marion, D.; Panda, S. Assessment of storm direct runoff and peak flow rates using improved SCS-CN models for selected forested watersheds in the Southeastern United States. *J. Hydrol. Reg. Stud.* **2020**, *27*, 1–18. [CrossRef]

53. Maidment, D.W.; Hoogerwerf, T.N. *Parameter Sensitivity in Hydrologic Modeling*. Technical Report; The University of Texas: Austin, TX, USA, 2002.

54. Wałęga, A. An attempt to establish regional dependecies for the parameter calculation of the Snyder's syntetic unit hydrograph. *Infrastruct. Ecol. Rural Areas* **2012**, *2*, 5–16.

55. Ignacio, J.A.F.; Cruz, G.T.; Nardi, F.; Henry, S. Assessing the effectiveness of a social vulnerability index in predicting heterogeneity in the impacts of natural hazards: Case study of the Tropical Storm Washi flood in the Philippines. *Vienna Yearb. Popul. Res.* **2015**, *13*, 91–129.

56. Cerdà, A.; Rodrigo-Comino, J.; Novara, A.; Brevik, E.C.; Vaezi, A.R.; Pulido, M.; Gimenez-Morera, A.; Keesstra, S.D. Long-term impact of rainfed agricultural land abandonment on soil erosion in the Western Mediterranean basin. *Prog. Phys. Geogr.* **2018**, *42*, 202–219.

57. Keesstra, S.D. Impact of natural reforestation on floodplain sedimentation in the Dragonja basin, SW Slovenia. *Earth Surf. Process. Landf.* **2007**, *32*, 49–65. [CrossRef]

58. Recanatesi, F.; Petroselli, A.; Ripa, M.N.; Leone, A. Assessment of stormwater runoff management practices and BMPs under soil sealing: A study case in a peri-urban watershed of the metropolitan area of Rome (Italy). *J. Environ. Manag.* **2017**, *201*, 6–18. [CrossRef]

 *water*

Article

A Calibrated, Watershed-Specific *SCS-CN* Method: Application to Wangjiaqiao Watershed in the Three Gorges Area, China

Lloyd Ling [1,*,†]**, Zulkifli Yusop** [2]**, Wun-She Yap** [1]**, Wei Lun Tan** [1]**, Ming Fai Chow** [3] **and Joan Lucille Ling** [4]

[1]   Universiti Tunku Abdul Rahman, Jalan Sungai Long, Bandar Sungai Long, Kajang 43000, Malaysia;
      yapws@utar.edu.my (W.-S.Y.); tanwl@utar.edu.my (W.L.T.)
[2]   Centre for Environmental Sustainability and Water Security, Universiti Teknologi Malaysia,
      Skudai 81310, Malaysia; zulyusop@utm.my
[3]   Institute of Energy Infrastructure, Universiti Tenaga Nasional, Kajang 43000, Malaysia;
      chowmf@uniten.edu.my
[4]   Department of Liberal Arts and Languages, American Degree Programme, Taylor's University,
      No. 1, Jalan Taylors, Subang Jaya 47500, Malaysia; linglucille@gmail.com
*   Correspondence: linglloyd@utar.edu.my; Tel.: +60-10-885-2735
†   Current address: Department of Civil Engineering, Lee Kong Chian Faculty of Engineering and Science,
      Universiti Tunku Abdul Rahman, Kajang 43000, Malaysia.

Received: 27 September 2019; Accepted: 26 November 2019; Published: 22 December 2019

Abstract: The Soil Conservation Service curve number (*SCS-CN*) method is one of the most popular methods used to compute runoff amount due to its few input parameters. However, recent studies challenged the inconsistent runoff results obtained by the method which set the initial abstraction ratio λ as 0.20. This paper developed a watershed-specific *SCS-CN* calibration method using non-parametric inferential statistics with rainfall–runoff data pairs. The proposed method first analyzed the data and generated confidence intervals to determine the optimum values for *SCS-CN* model calibration. Subsequently, the runoff depth and curve number were calculated. The proposed method outperformed the runoff prediction accuracy of the asymptotic curve number fitting method, linear regression model and the conventional *SCS-CN* model with the highest Nash–Sutcliffe index value of 0.825, the lowest residual sum of squares value of 133.04 and the lowest prediction error. It reduced the residual sum of squares by 66% and the model prediction errors by 96% when compared to the conventional *SCS-CN* model. The estimated curve number was 72.28, with the confidence interval ranging from 62.06 to 78.00 at a 0.01 confidence interval level for the Wangjiaqiao watershed in China.

Keywords: *SCS*; initial abstraction ratio; curve number; bootstrap; rainfall–runoff model

---

1. Introduction

Accurate direct surface runoff is essential for water resources' planning and development to reduce the occurrence of sedimentation and flooding at their downstream areas [1–3]. The simpler hydrological model under the law of parsimony with the least required input parameters and superior predictive model performance is preferable by many researchers [4–10].

Despite the existence of many rainfall–runoff models, the Soil Conservation Service curve number (*SCS-CN*) model proposed by the SCS National Engineering Handbook (SCS NEH) [11] is widely used in hydrological design [12,13]. The parameters of curve number (*CN*) and initial abstraction ratio (λ) are important in the *SCS-CN* method. The initial abstraction ratio (λ) plays a vital role in *SCS-CN* model in order to obtain an accurate runoff estimation [14]. Since 1954, λ was proposed by *SCS* to be

0.20. Equation (1) measures the runoff depth (Q) based on λ, the maximum potential water retention amount (S) and the rainfall depth (P). Throughout this paper, P, S and Q are measured in the unit of millimeters unless stated otherwise.

$$Q = \frac{(P - I_a)^2}{P - I_a + S}; P > I_a,$$ (1)

where I_a is the initial abstraction in the unit of millimeters computed using Equation (2).

$$I_a = \lambda S.$$ (2)

By substituting $\lambda = 0.20$ as proposed by *SCS*, Equation (3) is obtained.

$$Q = \frac{(P - 0.2S)^2}{P + 0.8S}.$$ (3)

On the other hand, the parameter CN is a transformation of S (where λ value must be 0.20), as shown in Equation (4).

$$CN = CN_{0.20} = \frac{1000}{10 + \dfrac{S}{25.4}}.$$ (4)

However, recent studies concluded that λ should not be a fixed value. Furthermore, some researchers also reported that λ value variation away from the proposed 0.20 value and lower than 0.20 was able to achieve better estimation of runoff prediction results in their studies [15–22]. Based on the median values of natural data for 307 watersheds, a group of researchers from the United States of America suggested a rounded value of $\lambda = 0.05$ to produce a better estimation of runoff depth [18]. Similarly, a group of researchers adopted the λ value of 0.05 for their research study at the Wangjiaqiao watershed in China [13].

The satellite imaging technique and geographic information system (GIS) were incorporated with the conventional *SCS-CN* method for studies but no attempt was reported to calibrate the primary *SCS-CN* rainfall–runoff framework with statistics in recent years. Recently, two groups of researchers developed a globally gridded CN dataset at 250 m spatial resolution [23,24]. However, the 250 m resolution dataset only represents general patterns of soil runoff potential appropriate for regional to global-scale analyses and may not capture the local variance suitable for fine-scale applications. Users need to check local conditions and runoff trends whenever available in their area of interest.

Since the tabulated NEH CN handbook [25] was based on the proposal that $\lambda = 0.20$, when λ changed, the CN value was altered and could not be determined or referred from the NEH handbook directly. Based on the study results of a group of researchers from the United States of America, $\lambda = 0.05$ was reported as the best value to represent watersheds in the United States of America and they proposed Equation (5) for runoff prediction while the S correlation equation between $S_{0.05}$ and $S_{0.20}$, as shown in Equation (6) (in inches), was used to transform the $S_{0.05}$ value back to $S_{0.20}$ in order to calculate CN values in their studies [18]. Without the correlation between $S_{0.05}$ and $S_{0.20}$, the direct substitution of S_λ (i.e., $S_{0.05}$) into Equation (4) yields CN_λ (the conjugate CN), which is totally different from the conventional CN (denoted as $CN_{0.20}$, where $\lambda = 0.20$) value [18].

$$Q = \frac{(P - 0.05S_{0.05})^2}{P + 0.95S_{0.05}}; P > 0.05S_{0.05}$$ (5)

$$S_{0.05} = 1.33(S_{0.20})^{1.15}$$ (6)

$$CN_{0.05} = \frac{100}{1.879[(100/CN_{0.20}) - 1]^{1.15} + 1}.$$ (7)

Since λ varies from location to location, a watershed-specific *SCS-CN* calibration method was proposed by using non-parametric inferential statistics based on rainfall and runoff data pairs. In this study, λ was no longer fixed at 0.20. This paper presents the use of inferential statistics to calibrate the primary *SCS-CN* rainfall–runoff model. To measure the effectiveness of the watershed-specific *SCS-CN* calibration method, the dataset from a past research in China was used to derive a watershed-specific *SCS-CN* rainfall–runoff model, a watershed-specific *S* correlation equation, λ value and the *CN* of Wangjiaqiao watershed to further improve their runoff prediction accuracy [13]. A calibrated, watershed-specific *SCS-CN* rainfall–runoff model was developed while an equation was derived to correct the runoff prediction of the conventional *SCS-CN* model. To date, no other published work has incorporated inferential statistics to calibrate the *SCS-CN* model.

2. The Proposed Calibrated Watershed-Specific *SCS-CN* Method

According to *SCS*, the initial abstraction (I_a) must be smaller than the smallest rainfall amount in the dataset which initiated runoff [25]. Furthermore, *SCS* constraint also stated that λ value must be in the range of [0, 1] and *S* must be a positive integer [25]. Based on Equation (3), the λ value cannot be a negative integer and *S* should be larger than I_a in order to meet its stated constraints.

The non-parametric inferential statistics of the bias corrected and accelerated (BCa) bootstrapping method [26–28] was conducted on the given dataset with 2000 random samples (with replacement) to make a statistically significant selection of key parameters—λ and *S*—with 99% confidence interval (CI) in order to calibrate Equation (1) [26,29]. The data distribution free, BCa bootstrapping technique was used in this study because it is robust and able to produce confidence intervals for statistical assessment.

The selection of mean or median λ and *S* is an universal dilemma in the hydrological field among researchers [18,30]. IBM statistical software SPSS (version 18.0) was used in this study, the normality test was conducted in SPSS to determine whether the optimum λ and *S* values should have been chosen from the mean or median confidence intervals. If a given dataset has less than 2000 samples, the Shapiro–Wilk test is suggested rather than the Kolmogorov–Smirnov test. This paper used a dataset which was less than 2000 samples, and therefore, the Shapiro–Wilk test was used. If the p value of the Shapiro–Wilk test is greater than 0.05, then the dataset is considered normally distributed [31]. As such, parameter optimization process should be inferred from the mean CI.

The supervised, non-linear genetic optimization algorithm was used in this study to search for the optimum λ and *S* values. The optimization algorithm created a population size of 2000, and 2000 random seeds with a mutation rate of 0.075 to converge towards an optimal solution within BCa 99% CI at a small error margin of 0.001 mm to search for the optimum λ and *S* value within the selected confidence interval range while the least square fitting algorithm minimized the residual sum of squares (*RSS*) between the predicted runoff and the values of the entire dataset.

As proposed by *SCS*, the *S* value was calculated from *CN* equation, as shown in Equation (4), whereas the *CN* value was chosen from the NEH handbook [25]. When λ is no longer equal to 0.20, a different λ value will yield a different *S* value, denoted as S_λ. In this study, Equation (1) was rearranged to illustrate a way to solve for S_λ, as shown in Equation (8):

$$S_\lambda = \frac{[P - \frac{(\lambda - 1)Q}{2\lambda}] - \sqrt{PQ - P^2 + [P - \frac{(\lambda - 1)Q}{2\lambda}]^2}}{\lambda}$$

$$Let : A = [P - \frac{(\lambda - 1)Q}{2\lambda}]$$

$$S_\lambda = \frac{A - \sqrt{PQ - P^2 + A^2}}{\lambda}. \tag{8}$$

Equation (8) is known as the general *S* equation denoted by S_λ. For the conventional *SCS-CN* model where $\lambda = 0.20$, $S_{0.20}$ can be calculated with Equation (8) according to the corresponding rainfall and runoff data pair. Since the optimum λ value of the calibrated model might be different from

$\lambda = 0.20$, a statistically significant S correlation is needed to correlate the S_λ to $S_{0.20}$ [18] in order to determine the equivalent $S_{0.20}$ value for the substitution back to Equation (4) to derive an equivalent $CN_{0.20}$ used by SCS practitioners. Without the S correlation equation, the CN value derived from any λ value which is not equal to 0.20 is known as the conjugate curve number denoted as CN_λ [18].

The watershed-specific rainfall–runoff model and the conventional SCS-CN model were derived from Equation (1); thus, the runoff prediction differences (Q_v) between two models can be modeled according to rainfall depth values in order to adjust the runoff prediction results of Equation (3) with a corrected equation.

The proposed method consists of two main steps: (1) Analyze the rainfall and runoff data pairs using the IBM statistical software SPSS (version 18.0) by generating confidence intervals for both mean and median values of derived λ and S; subsequently, perform a normality test to decide whether the confidence interval of mean or median value is to be used for optimization. (2) Optimize from the confidence interval range. In short, given rainfall–runoff data pairs (P_i, Q_i), I_{a_i}, S_i and λ_i for $i > 0$, the proposed watershed-specific SCS-CN calibration method consists of the following steps:

1. Perform bootstrap, BCa procedure and normality test in SPSS (version 18.0 or an equivalent statistics software) for (λ_i, S_i).
2. Check the normality test results of S_i to see whether it is normally distributed or not:

    (a) If yes, refer to the mean BCa confidence interval for S optimization.
    (b) Otherwise, refer to the median BCa confidence interval for S optimization.

3. Check the normality test results of λ_i to see whether it is normally distributed or not:

    (a) If yes, refer to the mean BCa confidence interval for λ optimization.
    (b) Otherwise, refer to the median BCa confidence interval for λ optimization.

4. Substitute the λ_{optimum} and S_{optimum} value into Equation (1) to form a calibrated SCS runoff model.
5. Given (P_i, Q_i) and λ_{optimum}, compute S_{λ_i} with Equation (8).
6. Given (P_i, Q_i) and $\lambda_{0.2}$, compute $S_{0.2_i}$ with Equation (8).
7. Correlate $S_{0.2_i}$ and S_{λ_i} to form a S correlation equation.
8. Substitute the S correlation equation into Equation (4) to derive $CN_{0.2}$.

Remark 1. *The I_a values, S values and I_a/S values used in this paper came from a study in China [13]. The I_a values used were estimated based on the comparison of the hydrograph with rainfall graph, and the methodology to derive the I_a and S value for each event is presented in the study. IBM SPSS version 18.0 was used to conduct all statistical analyses in this paper.*

3. Application to Wangjiaqiao Watershed in the Three Gorges Area, China

3.1. Study Site and Rainfall-Runoff Dataset

Figure 1 shows the study area of the Wangjiaqiao watershed which lies in Zigui County of Hubei Province, China. It is located at latitude 31°5′ N and 31°9′ N to longitude 110°40′ N and 110°43′ N. The total area of this study site is 1670 hectares and it is about 50 km northwest of the Three-Gorges Dam. Twenty nine rainfall–runoff data pairs were collected from 1994 to 1996 as shown in Table 1 [13].

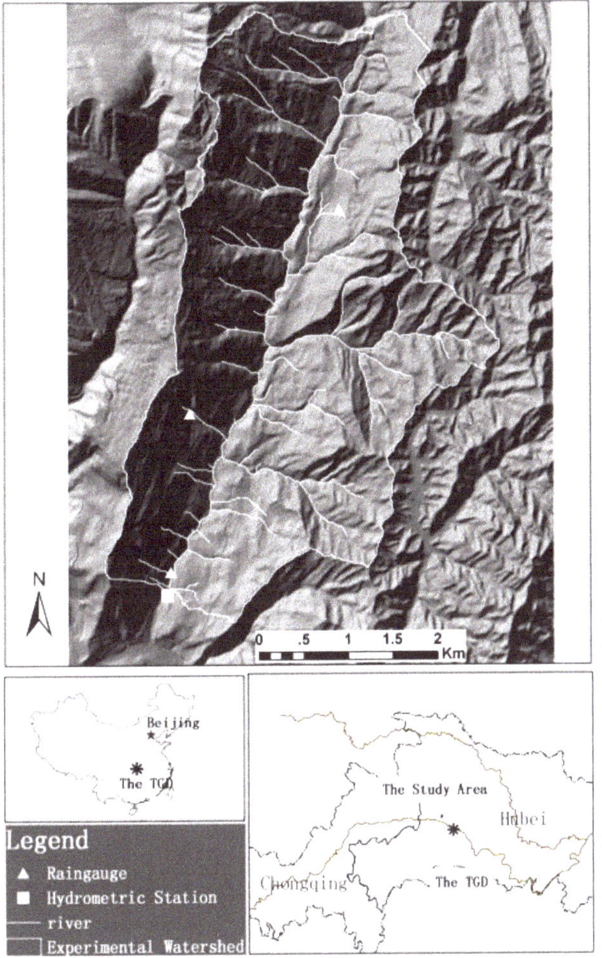

Figure 1. Location of study area in Wangjiaqiao watershed, China [13].

Table 1. Rainfall-runoff data of Wangjiaqiao watershed [13].

No.	Storm Event	Rainfall P (mm)	Direct Runoff Q (mm)	Initial Abstraction I_a (mm)	Retention S (mm)	Ratio I_a/S
1	03/05/1994	11.2	0.36	4.6	114.1	0.040
2	14/06/1995	11.9	0.16	6.5	176.9	0.037
3	24/04/1994	14.2	0.23	8.6	130.7	0.066
4	12/05/1995	19.0	0.01	16.8	481.8	0.032
5	16/04/1995	19.8	0.47	9.9	200.0	0.050
6	10/09/1995	22.0	0.03	18.1	503.1	0.036
7	01/06/1995	23.6	0.29	14.8	258.7	0.057
8	17/10/1995	24.1	0.48	13.6	218.0	0.062
9	09/05/1994	26.5	0.65	13.1	262.8	0.050
10	19/05/1995	27.1	1.78	10.0	146.9	0.068
11	19/06/1996	28.3	2.18	13.4	86.9	0.154
12	20/06/1995	30.8	4.02	8.3	103.4	0.080
13	29/07/1996	31.7	0.92	10.7	458.3	0.023
14	23/06/1996	32.3	1.27	6.1	514.7	0.012

Table 1. *Cont.*

No.	Storm Event	Rainfall P (mm)	Direct Runoff Q (mm)	Initial Abstraction I_a (mm)	Retention S (mm)	Ratio I_a/S
15	28/06/1996	32.4	1.75	14.6	163.3	0.089
16	14/05/1996	36.1	3.45	8.9	187.2	0.048
17	18/04/1994	38.1	3.72	10.4	178.6	0.058
18	19/10/1995	41.2	1.75	8.6	574.0	0.015
19	26/08/1994	48.1	1.08	19.0	755.2	0.025
20	04/06/1994	49.7	2.88	9.3	525.7	0.018
21	04/11/1996	49.8	4.15	6.7	405.1	0.017
22	07/07/1995	51.9	11.25	12.1	100.9	0.120
23	02/07/1996	54.0	2.86	7.4	714.0	0.010
24	02/05/1996	57.7	5.34	21.6	207.8	0.104
25	03/06/1996	62.1	3.23	18.5	544.6	0.034
26	07/06/1994	68.6	11.87	11.3	219.2	0.052
27	09/04/1994	73.7	9.94	14.1	297.6	0.047
28	18/09/1996	82.3	15.70	16.7	208.6	0.082
29	03/07/1996	85.9	21.31	7.8	208.1	0.037

3.2. Runoff Model Assessment

To measure the effectiveness of the proposed method, the Nash–Sutcliffe index (E), the model residual sum of square errors (RSS) and the overall model prediction error ($BIAS$) were computed using Equation (9)–(11) respectively.

$$E = 1 - \frac{RSS}{\sum_{i=1}^{n}(Q_{predicted} - Q_{mean})^2} \tag{9}$$

$$RSS = \sum_{i=1}^{n}(Q_{predicted} - Q_{observed})^2 \tag{10}$$

$$BIAS = \frac{\sum_{i=1}^{n}(Q_{predicted} - Q_{observerd})}{n}, \tag{11}$$

where n is the total rainfall–runoff events of this study.

RSS shows the model residual or prediction error. Thus, a predictive model with a lower RSS value is able to predict runoff amount better. Meanwhile, $BIAS$ shows the overall model prediction error by the summation of a model residual (prediction error). A predictive model with zero $BIAS$ value is able to achieve perfect runoff prediction results, whereas positive $BIAS$ value indicates the model tendency to over predict runoff amount and vice versa. Lastly, E index value is used to determine the model prediction efficiency of a model. E index ranges from $-\infty$ to 1.0, where 1.0 implies a perfectly predictive model [32]. E index values between 0 and 1 are generally viewed as acceptable levels of performance; however, when $E < 0$, the use of the mean runoff value observed can even predict the dataset better than the predictive model [33].

3.3. Results and Discussion

3.3.1. Inferential Statistics Assessment to Obtain Optimum λ and S

In total, 29 different pairs of λ and S values were derived from the rainfall–runoff data pairs of Wangjiaqiao watershed. At $\alpha = 0.01$ level, the proposed method searched for the optimum value within the CI range of mean and median λ values. Finally, an optimized pair of S and λ values was used to represent the watershed. The descriptive statistics of λ and S values were tabulated in Table 2.

Other than referring to the skewness and kurtosis values for λ and S dataset, the median value will be a better collective representation for λ and S dataset to represent the watershed, as the Shapiro–Wilk normality test also concluded that $p < 0.05$ for both λ and S dataset. As the data distribution for both λ and S dataset are not normally distributed by nature, the best collective λ and S values were optimized within the median confidence interval ranges of λ and S at $\alpha = 0.01$ level to minimize the RSS between the model predicted runoff amount and its observed values for the Wangjiaqiao dataset.

The optimum λ value was 0.043 while 260.081 mm was the optimum S value (denoted as: $S_{0.043}$). The product of the optimum S and λ value yield the initial abstraction (I_a) of 11.19 mm which was smaller than the smallest rainfall amount in the dataset from [13]. It fulfilled the SCS constraints, whereby the I_a amount must be met before any runoff process. Thus, based on the proposed watershed-specific SCS-CN calibration method, the runoff depth (Q) for the Wangjiaqiao watershed in China can be computed using Equation (12).

$$Q_{0.043} = \frac{(P - 11.19)^2}{P + 248.891}.$$
(12)

Based on Table 2, neither the mean nor the median BCa 99% λ CI includes the value of 0.20, and therefore, a λ value of 0.20 is not even statistically significant for the dataset of [13] at $\alpha = 0.01$ level. Furthermore, the standard deviation for λ at BCa 99% level is 0.034 with 77.14% λ fluctuation percentage between its lower and upper CI ranges to show that λ cannot be a constant but a variable due to its high fluctuation nature.

Table 2. Descriptive Statistical Results of λ and S at Bias Corrected and Accelerated (BCa) 99% Confidence Interval.

Wangjiaqiao Datasets	Descriptive Statistics of λ	BCa 99% Confidence Interval		Descriptive Statistics of S	BCa 99% Confidence Interval	
		Lower	Upper		Lower	Upper
Mean	0.053	0.069	0.384	308.477	230.413	395.527
Median	0.048	0.035	0.062	219.188	178.561	458.348
Skewness	1.251	0.268	1.846	0.867	−0.160	2.275
Kurtosis	1.884	−0.786	4.830	−0.389	−1.833	6.030
Std. Deviation	0.034	0.021	0.044	192.843	137.400	232.351

3.3.2. Watershed-Specific S Correlation Equation and CN for Wangjiaqiao Watershed in China

In the study of [13], they referred to the S correlation equation mapped by [18] where the median λ value of 0.05 was reported as the better collective representation for US watersheds. According to [18], a S correlation equation is required to convert the conjugate CN value when λ is no longer equal to 0.20. A different λ value will lead to different corresponding S value and the CN value will change accordingly. This study used Equation (8) to substitute $\lambda = 0.043$ and 0.20 with respective rainfall–runoff data pairs to calculate the corresponding $S_{0.043}$ and $S_{0.20}$ values in order to determine the S correlation equation between $S_{0.043}$ and $S_{0.20}$ values for Wangjiaqiao watershed in SPSS. Equation (6) should not be adopted as it was derived (in inches) to reflect watershed conditions of the United States of America [18].

SCS practitioner(s) will choose the CN value from the NEH handbook and calculate the S value with Equation (4) which was proposed by SCS under the hypothesis that $\lambda = 0.20$. This study used the reverse methodology to convert the $S_{0.043}$ value into an equivalent $S_{0.20}$ value through a mapped watershed-specific S correlation equation in order to determine the watershed-specific CN ($CN_{0.20}$). SPSS mapped the best S correlation equation between $S_{0.043}$ and $S_{0.20}$, as shown in Equation (13).

$$S_{0.2} = (S_{0.043})^{0.823}.$$
(13)

The correlation in Equation (13) has a lower standard error of 0.228 mm; the adjusted R^2 (Adj R^2) is equal to 0.998; and its p value is less than 0.001. As the optimum $S_{0.043} = 260.081$ mm, the equivalent $S_{0.20}$ value can be found by using Equation (13) with 97.39 mm, leading to the calculated watershed-specific CN value of 72.28 with Equation (4) for Wangjiaqiao watershed in China.

3.3.3. Asymptotic CN of Wangjiaqiao Watershed

Many studies concluded that the CN value could be derived with rainfall–runoff data pairs [34–36]. In 1993, the asymptotic CN fitting method (AFM) was introduced to determine the CN of a watershed with rainfall–runoff data pairs only [34]. At the WangJiaoQiao watershed, the standard CN behavior was detected with AFM, whereas CN stabilized at 65.10 (see Figure 2; hence, the equivalent $S_{0.2}$ value of the asymptotic CN can be calculated with Equation (4) as 136.17 mm).

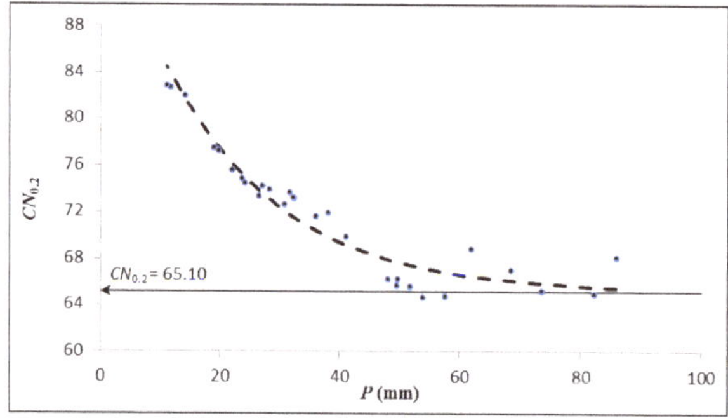

Figure 2. Standard behavior pattern, $CN_{\text{inf}} = 65$ for Wangjiaqiao watershed.

Furthermore, the I_a value can be determined, as it is the multiplication of λ and S values, and therefore, the I_a value we got was 27.24 mm. However, ten out of twenty nine (34.48%) rainfall events observed in Table 1 are smaller than the calculated I_a value. As such, AFM derived an I_a value which was in conflict with the aforementioned SCS constraint, one that meant there would be no runoff generated from any rainfall amount below the I_a value.

3.3.4. Residual Modeling and the Corrected Equation

Some researchers developed new models or modified the existing SCS-CN rainfall–runoff model by adding more parameters to improve surface runoff prediction accuracy [14,16,22,37–44]. However, those modified rainfall–runoff models could not solve the problem faced by SCS practitioners or any software that has already integrated the conventional SCS-CN model or embedded $\lambda = 0.20$ into its software algorithm.

In order to correct the runoff prediction variance (Q_v) between the conventional SCS-CN model and the new calibrated SCS-CN model to benefit SCS practitioners in their current practice, residual analyses of runoff predictive model were conducted between the two models to form a corrected equation, as shown in Equation (14). The Q_v was mapped with several non-linear regression models in SPSS according to rainfall values.

$$Q_v = -0.00003P^3 + 0.007P^2 - 0.322P + 3.511, \tag{14}$$

where Q_v is the runoff prediction difference (mm) obtained by computing Equations (3)–(12) and P is the rainfall depth (mm).

Equation (14) shows the Q_v between Equations (3) and (12). Positive Q_v indicates that Equation (3) predicted a larger runoff amount compared to Equation (12). Q_v can be plotted in order to visualize that Equation (3) produced inconsistent runoff prediction results, where it over predicted runoff when rainfall was less than 16 mm and more than 36 mm (see Figure 3) but under predicted the runoff amount when rainfall depth was between 16 mm and 36 mm. Equation (14) achieved an Adj R^2 near to 1.0 and low standard error of the estimate (0.053 mm) with statistical significance ($p < 0.001$) to correct and improve the runoff prediction results of Equation (3). As such, Equation (14) can be amended to Equation (3) to improve the runoff prediction accuracy of Equation (3), as shown in Equation (15).

$$Q = \frac{(P - 0.2S)^2}{P + 0.8S} - (-0.00003P^3 + 0.007P^2 - 0.322P + 3.511),\tag{15}$$

where $P > 0.2S$, else $Q = 0$.

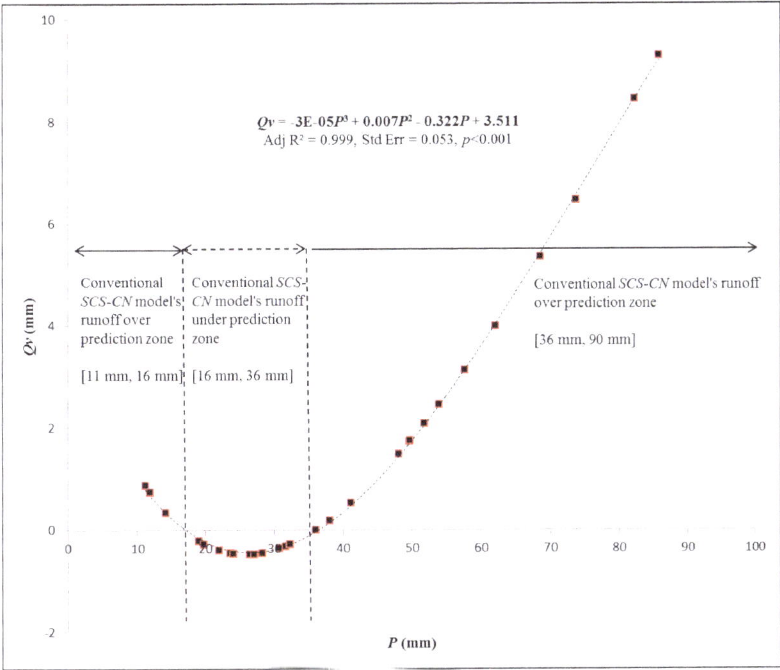

Figure 3. Runoff difference between the proposed calibrated model and the conventional *SCS-CN* model.

Equation (15) adjusted and improved the runoff prediction results of Equation (3). *RSS* of Equation (3) was reduced by 66%; the over prediction tendency of the model was corrected by 96%; and achieved proximate runoff prediction results as the new calibrated *SCS-CN* model, while the *E* index was improved by 71% to 0.826. Without model calibration, the conventional *SCS-CN* model over-predicted runoff amount by almost 155,000 m³ at the rainfall depth of 85.90 mm when compared to the newly calibrated *SCS-CN* model at the 1670 ha Wangjiaqiao watershed in China. The runoff over prediction risk would be even worse toward high rainfall intensities. On the other hand, it also under-predicted runoff amount up to 8000 m³ at the rainfall depth of 26.5 mm. This showed that the conventional *SCS-CN* model was not only statistically insignificant at $\alpha = 0.01$ level, but produced inconsistent runoff prediction results at different rainfall depths.

3.3.5. Comparison of Runoff Prediction Models

The law of parsimony favors a simple model with less fitting parameters. As such, this study explored the possibility of using a linear regressed rainfall–runoff model to quantify the runoff behavior at the Wangjiaqiao watershed. Past researchers proposed that the slope of a linear fitting equation could represent the total impervious area of a watershed while the fitting constant was regarded as the depression loss [45]. In this study, SPSS fitted the best linear regressed rainfall–runoff model for Wangjiaqiao watershed as: $Q = 0.214P - 4.623$ with an Adj $R^2 = 0.715$ and standard error of estimate was equal to 2.779 mm. Both fitting slope and constant parameter were statistically significant ($p < 0.001$) but the model produced five out of twenty nine (17.2%) negative runoff prediction results.

λ's confidence interval in Table 2 implies that λ value cannot be 0.20 and a constant value at Wangjiaqiao watershed in China. As a result, the conventional *SCS-CN* model becomes invalid and not statistically significant. When the λ value is fixed at 0.20, the optimized $S_{0.20} = 100.8$ mm and I_a value can be calculated as 20.16 mm, but this I_a value violated the *SCS* constraint, as 17.24% of rainfall data pairs from [13] were less than the I_a value. Thus, the conventional *SCS-CN* model faced the same problem as the AFM and the linear regression model. Moreover, the conventional *SCS-CN* model had the lowest *E* index; highest *RSS*; and *BIAS* when compared to the AFM, the linear regression model and the newly calibrated watershed-specific *SCS-CN* model. The statistics of five runoff predictive models were tabulated in Table 3.

Table 3. Descriptive statistics of five runoff predictive models

Model Parameters and Statistics	AFM Model	Calibrated SCS-CN Model Equation (12)	Conventional SCS-CN Model Equation (3)	Corrected SCS-CN Model Equation (15)	Linear Regression Model
p value	-	<0.001	Not Significant	Adjusted	<0.001
λ	0.200	0.043	0.200	0.200	-
S (mm)	106.19	200.08	100.80	-	-
I_a (mm)	27.238	11.190	20.16	-	4.623
E	0.799	0.825	0.482	0.826	0.725
RSS	152.251	133.044	393.126	131.960	208.462
$BIAS$	−0.624	0.056	1.586	0.055	−0.008
$CN_{0.2}$	65.096	72.284	71.590	-	-
Residual Mean	−0.624	0.056	n/a	0.055	−0.008
BCa 99% CI Range of Mean Residual	[−1.631, 0.283]	[−0.898, 0.961]	n/a	[−0.896, 0.970]	[−1.225, 1.124]
Residual Median	n/a	n/a	0.140	n/a	n/a
BCa 99% CI Range of Median Residual	n/a	n/a	[−0.420, 3.850]	n/a	n/a
Standard Deviation of Model error	2.244	2.179	3.381	2.171	2.728
Variance (Residual)	5.035	4.748	11.429	4.715	7.445
Range	11.350	10.844	12.740	10.770	12.987
p value of Shapiro-Wilk Test	0.197	0.111	0.012	0.121	0.597

Residual analyses were conducted by using SPSS to measure the runoff prediction error of every runoff predictive model. The model with the smallest residual confidence interval range, lowest standard deviation error and variance was to be the best runoff predictive model of this study. The SPSS normality test showed that the significant value of the Shapiro–Wilk test was more than 0.05 for the AFM model, the newly calibrated watershed-specific *SCS-CN* model, the corrected *SCS-CN* model and the linear regression model; thus, their mean residual values were referred to for the accuracy comparison of the predictive model. On the other hand, the p value of the Shapiro–Wilk test for the conventional *SCS-CN* model was less than 0.05; hence, its median residual values were used as the benchmark for its model accuracy.

The mean residual value of the newly calibrated, watershed-specific *SCS-CN* model was among the lowest (0.056 mm) and nearest to zero, while its 99% BCa confidence interval range of mean residuals spanned across a small range when compared to other models. In addition, the newly calibrated, watershed-specific *SCS-CN* model had a low residual variance and standard deviation which indicated that the newly calibrated runoff predictive model had the ability to achieve a runoff prediction with low error. As a result, the newly calibrated watershed-specific *SCS-CN* model became a suitable runoff predictive model for the twenty nine data pairs at the Wangjiaqiao watershed in this study. On the other hand, the corrected *SCS-CN* model also managed to correct runoff prediction errors of the conventional *SCS-CN* model and achieved proximate runoff prediction results as the newly calibrated watershed-specific *SCS-CN* model, which proves that the presented residual modeling technique was effective at transforming Equation (3) into a better rainfall–runoff model.

Without model calibration, the conventional *SCS-CN* model over predicted runoff volume significantly from rainfall depths of 40 mm onward at the 1670 ha Wangjiaqiao watershed in China (see Figure 4). Through Equation (14), it is possible to quantify and model the runoff over prediction volume according to its corresponding rainfall depths. SPSS mapped that Q_v (m^3) $= 21.01P^2 + 642.62P - 54,520$ was able to quantify the runoff over prediction volume from the conventional *SCS-CN* model with and an Adj R^2 near to 1.0 and low standard error of 712.92 m^3 ($p < 0.001$).

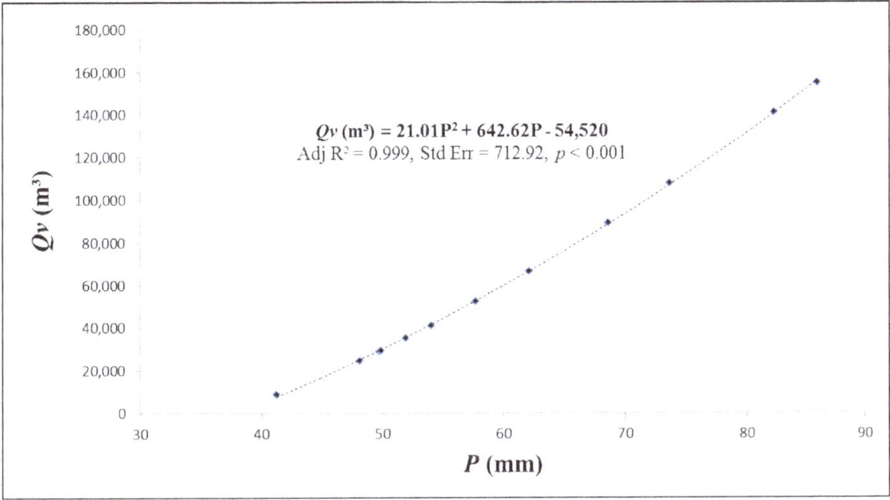

Figure 4. Runoff volumetric differences between the proposed, calibrated *SCS CN* model and the conventional *SCS-CN* model.

4. Conclusions

A new watershed-specific *SCS-CN* calibration method was proposed to identify the optimum λ and S values, and derive CN with the use of non-parametric inferential statistics, the rainfall-runoff data pairs and the supervised non-linear numerical optimization technique. The proposed model was applied to the Wangjiaqiao watershed in China. Inferential statistics showed that the conventional *SCS-CN* model was not statistically significant at $\alpha = 0.01$ level, and therefore, it was not applicable to model the runoff conditions of the Wangjiaqiao watershed in China. The proposed model identified the optimum median λ of 0.043 (with the 99% confidence interval ranging from 0.035 to 0.062) as the best collective λ for the Wangjiaqiao watershed. A watershed-specific S correlation equation was mapped in this study to show that the CN value of the Wangjiaqiao watershed can be derived directly without referring to the NEH handbook. The estimated CN of Wangjiaqiao watershed in China was 72.28 with a 99% confidence interval ranging from 62.06 to 78.0.

The newly calibrated watershed-specific *SCS-CN* model improved the previous study results: the E index increased by 7.4%, the *BIAS* of predictive model was reduced by 93.8% and model's *RSS* was lowered by 24.4%. These improvements were achieved with a λ of 0.043 instead of rounding it to 0.050. The proposed model also outperformed the AFM model, the conventional *SCS-CN* model and the linear regression model to predict runoff amount at the Wangjiaqiao watershed. The proposed model had the lowest *BIAS* and *RSS*, and the highest E index when compared to those runoff predictive models. On the other hand, the linear regression model had the second highest model inaccuracy after the conventional *SCS-CN* model, and both models produced negative runoff prediction results that were unable to yield a meaningful hydrological interpretation to predict surface runoff at the Wangjiaqiao watershed. Both models were also unable to produce positive runoff prediction results for rainfall depths less than 20 mm.

A runoff corrected equation was formulated through the proposed residual modeling technique under this study. The equation managed to correct runoff inconsistencies of the conventional *SCS-CN* model and improved its runoff prediction accuracy. The *S* correlation equation from other study cannot be adopted as it reflects specific watershed conditions. It must be derived with watershed-specific λ and rainfall–runoff data pairs in order to convert the CN_λ into an equivalent $CN_{0.20}$. This study also found that the rounding of λ and *CN* values will induce the runoff prediction errors. As a result, *CN* with at least two decimal places is recommended to *SCS* practitioners for their future studies.

Based on the proposed *SCS-CN* calibration method, Equation (12) is recommended for the runoff prediction of the dataset from [13] at the Wangjiaqiao watershed in the Three Gorges Area. When a new rainfall–runoff dataset becomes available, *SCS* practitioners should re-derive the calibrated *SCS* runoff model again with the proposed methodology.

Author Contributions: conceptualization and methodology, L.L.; software, W.-S.Y and J.L.; validation, Z.Y.; formal analysis, L.L.; investigation, W.L.L., M.F.C. and Z.Y.; resources, L.L.; writing—original draft preparation, L.L.; writing—review and editing, W.-S.Y., L.L. and J.L.L.; visualization, L.L. and J.L.L.; supervision, L.L. and Z.Y.; project administration, L.L.; funding acquisition, L.L. and Z.Y. All authors have read and agreed to the published version of the manuscript.

Funding: The authors would like to acknowledge the financial support received from the Universiti Tunku Abdul Rahman under the research grant:/RMC/UTARRF/2016-C2/L13.

Acknowledgments: The authors also appreciate the guidance from Richard H. Hawkins (University of Arizona, USA) and Wen Jia Tan (Universiti Tunku Abdul Rahman) to assist in some analyses.

Conflicts of Interest: The authors declare no conflict of interest.

Abbreviations

The following abbreviations are used in this manuscript:

SCS-CN	Soil Conservation Service Curve Number
NEH	National Engineering Handbook
CN	Curve Number
λ	Initial Abstraction Ratio
Q	Runoff Depth
S	Maximum potential water retention amount
P	The rainfall depth
BCa	Bias corrected and accelerated
S_λ	*S* value of different λ
CN_λ	Conjugate Curve Number
Q_v	Runoff prediction differences
E	Nash-Sutcliffe index
RSS	Model residual sum of square errors
BIAS	Overall model prediction error
CI	Confidence Interval
SPSS	IBM statistical software SPSS
Adj R^2	Adjusted R^2
AFM	Asymptotic *CN* fitting method

References

1. Wang, Z.; Jiao, J.; Rayburg, S.; Wang, Q.; Su, Y. Soil erosion resistance of "Grain for Green" vegetation types under extreme rainfall conditions on the Loess Plateau, China. *Catena* **2016**, *141*, 109–116. [CrossRef]
2. Zhou, J.; Fu, B.; Gao, G.; Lü, Y.; Liu, Y.; Lü, N.; Wang, S. Effects of precipitation and restoration vegetation on soil erosion in a semi-arid environment in the Loess Plateau, China. *Catena* **2016**, *137*, 1–11. [CrossRef]

3. Chen, H.; Zhang, X.P.; Abla, M.; Lu, D.; Yan, R.; Ren, Q.F.; Ren, Z.Y.; Yang, Y.H.; Zhao,W.H.; Lin, P.F. Effects of vegetation and rainfall types on surface runoff and soil erosion on steep slopes on the Loess Plateau, China. *Catena* **2018**, *170*, 141–149. [CrossRef]

4. Merritt, W.S.; Letcher, R.A.; Jakeman, A.J. A review of erosion and sediment transport models. *Environ. Model. Softw.* **2003**, *18*, 761–799. [CrossRef]

5. Beven, K. Changing ideas in hydrology—The case of physically based models. *J. Hydrol.* **1989**, *105*, 157–172. [CrossRef]

6. Ferro, V.; Minacapilli, M. Sediment delivery processes at basin scale. *Hydrol. Sci. J.* **1995**, *40*, 703–717. [CrossRef]

7. Loague, K.; Freeze, R.A. Comparison of rainfall-runoff modeling techniques on small upland catchments. *Water Resour. Res.* **1985**, *21*, 229–248. [CrossRef]

8. Perrin, C.; Michel, C.; Andreassian, V. Does a large number of parameters enhance model performance? Comparative assessment of common catchment model structures on 429 catchments. *J. Hydrol.* **2001**, *242*, 275–301. [CrossRef]

9. Steefel, C.I.; Van Cappellan, P. Reactive transport modeling of natural systems. *J. Hydrol.* **1998**, *209*, 1–7.

10. Kleissen, F.; Beck, M.; Wheater, H. The identifiability of conceptual hydrochemical models. *Water Resour. Res.* **1990**, *26*, 2979–2992. [CrossRef]

11. SCS National Engineering Handbook (SCS NEH). Section 4, Hydrology, Chapter 21. In *National Engineering Handbook*; SCS: Washington, DC, USA, 1965.

12. Shi, P.J.; Yuan, Y.; Zheng, J.; Wang, J.A.; Ge, Y.; Qiu, G.Y. The effect of land use/cover change on surface runoff in Shenzhen region, China. *Catena* **2007**, *69*, 31–35. [CrossRef]

13. Shi, Z.H.; Chen, L.D.; Fang, N.F.; Qin, D.F.; Cai, C.F. Research on the SCS-CN initial abstraction ratio using rainfall- runoff event analysis in the Three Gorges Area, China. *Catena* **2009**, *77*, 1–7. [CrossRef]

14. Jain, M.K.; Mishra S.K.; Babu S.; Singh, V.P. Enhanced runoff curve number model incorporating storm duration and a non-linear Ia-S relation. *J. Hydrol. Eng. ASCE* **2006**, *11*, 631–635. [CrossRef]

15. Ponce, V.M.; Hawkins, R.H. Runoff curve number: Has it reached maturity? *J. Hydrol. Eng.* **1996**, *1*, 11–19. [CrossRef]

16. Mishra, S.K.; Singh, V.P. Long-term hydrological simulation based on the soil conservation service curve number. *Hydrol. Process* **2004**, *18*, 1291–1313. [CrossRef]

17. Baltas, E.A.; Dervos, N.A.; Mimikou, M.A. Technical note: Determination of the SCS initial abstraction ratio in an experimental watershed in Greece. *Hydrol. Earth Syst. Sci.* **2007**, *11*, 1825–1829. [CrossRef]

18. Hawkins, R.H.; Ward, T.J.; Woodward, D.E.; Van Mullem, J.A. *Curve Number Hydrology: State of Practice*; American Society of Civil Engineers: Reston, VI, USA, 2009.

19. Fu, S.; Zhang, G.; Wang, N.; Luo, L. Initial abstraction ratio in the SCS-CN method in the Loess Plateau of China. *Trans. ASABE* **2011**, *54*, 163–169. [CrossRef]

20. D'Asaro, F.; Grillone, G. Empirical investigation of curve number method parameters in the Mediterranean area. *J. Hydrol. Eng* **2012**, *17*, 1141–1152. [CrossRef]

21. D'Asaro, F.; Grillone, G.; Hawkins, R.H. Curve number: Empirical evaluation and comparison with curve number handbook tables in Sicily. *J. Hydrol. Eng.* **2014**, *19*. [CrossRef]

22. Singh, P.K.; Mishra, S.K.; Berndtsson, R.; Jain, M.K.; Pandey, R.P. Development of a modified SMA based MSCS-CN model for runoff estimation. *Water Resour. Manag.* **2015**, *29*, 4111–4127. [CrossRef]

23. Ross, C.; Prihodko, L.; Anchang, J.; Kumar, S.; Ji, W.; Hanan, N. HYSOGs250m, global gridded hydrologic soil groups for curve-number-based runoff modeling. *Sci. Data* **2018**, *5*, 1–8. [CrossRef] [PubMed]

24. Jaafar, H.; Ahmad, F.; El Beyrouthy, N. GCN250, new global gridded curve numbers for hydrologic modeling and design. *Sci. Data* **2019**, *6*, 1–8. [CrossRef] [PubMed]

25. National Engineering Handbook (NEH). Part 630 Hydrology. In *National Engineering Handbook*; USDA: Washington, DC, USA, 2004.

26. Efron, B.; Tibshirani, R. *An Introduction to the Bootstrap*; Chapman & Hall/CRC: Boca Raton, FL, USA, 1994, ISBN 978-0-412-04231-7.

27. Davison, A.C.; Hinkley, D.V. Bootstrap methods and their application. In *Cambridge Series in Statistical and Probabilistic Mathematics*; Cambridge University Press: Cambridge, UK, 1997. ISBN 0-521-57391-2.

28. Efron, B. *Large-Scale Inference: Empirical Bayes Methods for Estimation, Testing and Prediction*; Institute of Mathematical Statistics Monographs, Cambridge University Press: Cambridge, UK, 2010, ISBN 978-0-521-19249-1.

29. Rochoxicz, J.A., Jr. Bootstrapping Analysis, Inferential Statistics and EXCEL. *Spreadsheets Educ.* **2011**, *4*, 4.

30. Schneider, L.E.; McCuen, R.H. Statistical guidelines for curve number generation. *J. Irrig. Drain. Eng.* **2005**, *131*, 282–290. [CrossRef]

31. Andy, F. *Discovering Statistics Using IBM SPSS*, 4th ed.; SAGE Publications: Thousand Oaks, CA, USA, 2013; pp. 595–609.

32. Nash, J.; Sutcliffe, J. River flow forecasting through conceptual models part 1 - A discussion of principles. *J. Hydrol.* **1970**, *10*, 282–290. [CrossRef]

33. Moriasi, D.N.; Arnold, J.G.; Van Liew, M.W.; Bingner, R.L.; Harmel, R.D.; Veith, T.L. Model evaluation guidelines for systematic quantification of accuracy in watershed simulations. *Trans. ASABE* **2007**, *50*, 885–900. [CrossRef]

34. Hawkins, R.H. Asymptotic determination of runoff curve numbers from data. *J. Irrig. Drain. Eng.* **1993**, *119*, 334–345. [CrossRef]

35. Soulis, K.X.; Valiantzas, J.D.; Dercas, N.; Londra, P.A. Investigation of the direct runoff generation mechanism for the analysis of the SCS-CN method applicability to a partial area experimental watershed. *Hydrol. Earth Syst. Sci.* **2009**, *13*, 605–615. [CrossRef]

36. Soulis, K.X.; Valiantzas, J.D. Identification of the SCS-CN parameter spatial distribution using rainfall-runoff data in heterogeneous watersheds. *Water Resour. Manag.* **2013**, *27*, 1737–1749. [CrossRef]

37. Mishra, S.K.; Singh, V.P. SCS CN method. Part-I. Derivation of SCS-CN based models. *Acta Geophy. Pol.* **2002**, *50*, 457–477.

38. Mishra, S.K.; Singh, V.P. *Soil Conservation Service Curve Number (SCS-CN) Methodology*; Kluwer: Dordrecht, The Netherlands, 2003; ISBN 1-4020-1132-6.

39. Sahu, R.K.; Mishra, S.K.; Eldho, T.I.; Jain, M.K. An advanced soil moisture accounting procedure for SCS curve number method. *Hydrol. Process.* **2007**, *21*, 2872–2881. [CrossRef]

40. Kim, N.W.; Lee, J. Temporally weighted average curve number method for daily runoff simulation. *Hydrol. Process.* **2008**, *22*, 4936–4948. [CrossRef]

41. Wang, S.; Kang, S.; Zhang, L.; Li, F. Simulation of an agricultural watershed using an improved curve number method in SWAT. *Trans. ASABE* **2008**, *51*, 1323–1339. [CrossRef]

42. Ajmal, M.; Moon, G.; Ahn, J.; Kim, T. Investigation of SCS-CN and its inspired modified models for runoff estimation in South Korean watersheds. *J. Hydro-Environ. Res.* **2015**, *9*, 592–603. [CrossRef]

43. Ajmal, M.; Waseem, M.; Wi, S.; Kim, T. Evolution of a parsimonious rainfall–runoff model using soil moisture proxies. *J. Hydrol.* **2015**, *530*, 623–633. [CrossRef]

44. Ajmal, M.; Khan, T.A.; Kim, T.W. A CN-based ensembled hydrological model for enhanced watershed runoff prediction. *Water* **2016**, *8*, 20. [CrossRef]

45. Arnbjerg-Nielsen, K.; Harremoes, P. Prediction of hydrological reduction factor and initial loss in urban surface runoff from small ungauged catchments. *Atmos. Res.* **1996**, *42*, 137–147. [CrossRef]

Article

Application of the SCS–CN Method to the Hancheon Basin on the Volcanic Jeju Island, Korea

Minseok Kang and Chulsang Yoo *

School of Civil, Environmental and Architectural Engineering, College of Engineering, Korea University, Seoul 02841, Korea; minseok0517@hanmail.net
* Correspondence: envchul@korea.ac.kr; Tel.: +82-10-9326-9168

Received: 8 October 2020; Accepted: 25 November 2020; Published: 29 November 2020

Abstract: This study investigates three issues regarding the application of the SCS–CN (Soil Conservation Service–Curve Number) method to a basin on the volcanic Jeju Island, Korea. The first issue is the possible relation between the initial abstraction and the maximum potential retention. The second is the determination of the maximum potential retention, which is also closely related to the estimation of CN. The third issue is the effect of the antecedent soil moisture condition (AMC) on the initial abstraction, maximum potential retention and CN. All of these issues are dealt with based on the analysis of several rainfall events observed in the Hancheon basin, a typical basin on Jeju Island. In summary, the results are that, firstly, estimates of initial abstraction, ratio λ, maximum potential retention, and CN were all found to be consistent with the SCS–CN model structure. That is, CN and the maximum potential retention showed a strong negative correlation, and the ratio λ and the maximum potential retention also showed a rather weak negative correlation. On the other hand, a significant positive correlation was found between CN and the ratio λ. Second, in the case where the accumulated number of days is four or five, the effect of antecedent precipitation amount is clear. The antecedent five-day rainfall amount for the AMC-III condition is higher than 400 mm, compared to the AMC-I condition of less than 100 mm. Third, an inverse proportional relationship is found between the AMC and the maximum potential retention. On the other hand, a clear linear proportional relation is found between the AMC and CN. Finally, the maximum potential retention for the Hancheon basin is around 200 mm, with the corresponding CN being around 65. The ratio between the initial abstraction and the maximum potential retention is around 0.3. Even though these results are derived by analyzing a limited number of rainfall events, they are believed to properly consider the soil characteristics of Jeju Island.

Keywords: SCS–CN method; Jeju Island; initial abstraction; maximum potential retention; antecedent moisture condition

---

1. Introduction

Jeju Island, formed by volcanic activity, has totally different hydrological characteristics from those of the inland area of mainland Korea [1]. Basalt, a volcanic rock, shows high permeability, and thus low surface runoff [2]. This is the reason why most channels on Jeju Island are dry, and runoff occurs only when the rainfall intensity is high [3]. According to the Jeju Special Self-Governing Province [4], the mean runoff ratio is just 21%, which is far lower than that in the inland mainland area. Different from the slowly increasing runoff pattern in the inland mainland area, the runoff on Jeju Island shows an abruptly rising pattern after the soil layer is fully saturated [5].

Several problems are raised regarding the rainfall–runoff analysis of Jeju Island. In fact, these problems are mostly related to the rainfall loss by infiltration. The Soil Conservation Service–Curve Number (SCS–CN) method is a common method adopted by most rainfall–runoff analyses in Korea. However, application of the SCS–CN method to Jeju Island has always been problematic. First, it is not easy to determine the maximum potential retention. The problem is simply to determine the curve number (i.e., CN). Determination of CN is also closely related to the classification of the hydrological soil group for the given soil or soil series. There are five different classification rules in Korea, all of which derive different classification results, and thus different CNs. The second problem is related to the amount of initial abstraction. Generally, the amount of initial abstraction is assumed to be 20% of the maximum potential retention [6]. Although the Guideline for Design Flood Estimation in Korea [7] proposed the use of 40% for Jeju Island instead of 20%, this guideline is relied on only by a few case studies [8]. It is also important to consider that this ratio is dependent upon CN. Additionally, the impact of the antecedent moisture condition (AMC) is also ambiguous. High porous basalt soil infiltrates so fast that simply applying the AMC condition that is applied to the inland mainland area of Korea does not work for Jeju Island.

There have been many attempts to solve these issues. First, Ha et al. [9] showed the typical application procedure of the SCS–CN method. In the analysis of the variation of the effective rainfall due to the change of land use on Jeju Island, they estimated CN by applying the classification rule of hydrological soil group proposed by the Rural Development Administration (RDA) [10], the initial abstraction being 20% of the maximum potential retention. The AMC was determined by considering the antecedent five-day precipitation. However, in the study of Jung [11], even though the same classification rule of the RDA [10] was used, the estimated CN was additionally modified by considering the basin slope. The formula by Sharpley and Williams [12] was applied for this purpose, and, as a result, CN was increased by more than 10%. Also, they suggested that the initial abstraction should be 35–45% of the maximum potential retention depending on the basin. The application procedure of the SCS–CN method of Jung et al. [13] was similar to that of Jung [11]. However, to consider the porous volcanic soil of Jeju Island, they proposed the use of the intermediate value of CNs for AMC-II and AMC-III condition, even though the antecedent five-day precipitation was very high. Recently, however, Yang et al. [14] returned to using the typical procedure of the SCS–CN method. That is, the initial abstraction was assumed to be 20% of the maximum potential retention, and CN was determined by considering the AMC condition. CN was increased by considering the basin slope. As can be found in the above mentioned studies, the application of the SCS–CN method on Jeju Island so far has not been consistent.

Several cases can also be found of these issues in volcanic regions worldwide. Interestingly, their results are not consistent either. For example, Leta et al. [15] showed that 40% of the maximum potential retention was a good ratio for the initial abstraction in the Heeia basin, Hawaii. Most of the Heeia basin is covered by volcanic silty clay soils, and CN was also estimated to be very small at around 35–62 under normal condition (i.e., AMC-II condition). The Bathan basin in Saudi Arabia shows a different result [16]. Even though the Bathan basin lies in an arid volcanic region covered by basalt soil, the initial abstraction was estimated to be 20% of the maximum potential retention. CN was estimated very high at 78.5. In the application case of Kyushu Island in Japan, the initial abstraction was assumed to be 20% of the maximum potential retention, but CN was less than 70 [17]. In the application to the Naivasha basin in Kenya, the initial abstraction was assumed to be 20% of the maximum potential retention, but CN was very small at just 61 [18]. However, in the application to the La Lumbre and Montegrande basins in the Volcán de Colima, Mexico, the initial abstraction was assumed to be 20% of the maximum potential retention, but CN was very high at 75.5 [19].

The objective of this study is to investigate the above-mentioned issues of the SCS–CN method raised in application to the volcanic Jeju Island. First, this study investigates the possible relation between the initial abstraction and the maximum potential retention. Second, the proper value of the maximum potential retention of Jeju Island is estimated. As mentioned before, this issue is closely related to the estimation of CN, which is also dependent upon the classification rule of the hydrological soil group of soil or soil series. Finally, the effect of the AMC on CN is the third issue of interest. Because the soil on Jeju Island is porous and highly permeable, the typical AMC conditions used in the inland mainland area of Korea may not be applicable. In this study, these issues are dealt with based on the analysis of several rainfall events observed in the Hancheon basin. The Hancheon basin is a highly typical basin on Jeju Island, so the results derived in the analysis may be applicable to the other basins.

2. Theoretical Backgrounds

2.1. The SCS–CN Method

The SCS–CN method was developed by the U.S. Soil Conservation Service [6] and considers the soil, land use, vegetation, etc. to derive the curve number (CN) as a representative parameter for the hydrologic infiltration characteristics. CN ranges from 0 to 100.

The SCS–CN method assumes that the ratio between the actual retention (F) and the maximum potential retention (S) is the same as that between the direct runoff (Q) and the total precipitation (P). Here, the actual retention represents the precipitation portion that is not converted into the direct runoff, even after the runoff begins, i.e., $F = P - Q$:

$$\frac{Q}{P} = \frac{F}{S} \tag{1}$$

However, in actual application, the above equations are modified somewhat by considering the initial abstraction (I_a). Here, the initial abstraction is introduced to represent various losses, like interception and evaporation. As a result, the above equation becomes valid by replacing P by $(P - I_a)$:

$$\frac{Q}{P - I_a} = \frac{F}{S} \tag{2}$$

Now the above Equation (2) can be arranged to represent the direct runoff:

$$Q \begin{aligned} &= \frac{(P-I_a)^2}{(P-I_a)+S} \quad (P > I_a) \\ &= 0 \quad\quad\quad\quad\ (P \leq I_a) \end{aligned} \tag{3}$$

The direct runoff, expressed as a function of I_a and S, is now simplified once more by introducing a linear relation $I_a = \lambda S$ [6]:

$$Q = \frac{(P - \lambda S)^2}{P + (1 - \lambda)S} \tag{4}$$

The ratio λ is generally assumed to be 0.2 [6].

With a given maximum potential retention, the direct runoff can be estimated from the total precipitation. The SCS also proposed using the CN to estimate the maximum potential retention. That is:

$$S = \frac{25,400}{CN} - 254 \tag{5}$$

As can be expected, in the case that CN is zero, the maximum potential retention becomes infinite. On the other hand, a CN of 100 indicates that the impervious area needed for maximum potential retention becomes zero (i.e., $Q = P$).

The initial abstraction is a key factor in determining the initiation of the direct runoff. However, its estimation involves much uncertainty, which results in error in the estimation of the direct runoff [6]. Especially in a large basin when the areal-average rainfall and runoff data are the only available information, the estimation of the initial abstraction becomes ambiguous. As mentioned earlier, the common assumption of the initial abstraction is 20% of the maximum potential retention [6]. However, many studies have disputed this linear assumption between the initial abstraction and maximum potential retention [20–24]. Another study proposed the range of 0–40% as the ratio between the initial abstraction and maximum potential retention [25]. Other studies, like that of Woodward et al. [20], showed how to determine the initial abstraction by analyzing the rainfall–runoff data. In fact, they defined the initial abstraction to be the total rainfall amount up to the starting point of the direct runoff. Based on the analysis of 134 basins in the US (United States), Woodward et al. [20] proposed just 5% of the maximum potential retention for the initial abstraction.

The maximum potential retention is the maximum volume of water to be stored within the basin soil profile. In early application of the SCS–CN method, the maximum potential retention was assumed to be constant for the given basin [26]. Theoretically, the maximum potential retention can be the volume of void in the soil profile, but it is also dependent upon the antecedent moisture condition. This is simply because recent precipitation can still remain within the void of the soil profile [25,27–30]. The maximum potential retention also varies depending on the vegetation, land use, antecedent precipitation, etc. [31]. As the maximum potential retention is defined as the maximum difference between the total precipitation and the direct runoff, it should be determined by analyzing their relation [30].

The antecedent moisture condition (AMC) thus plays an important role in the application of the SCS–CN method. The AMC simply defines the soil condition to control the infiltration capacity [31]. Three different concepts are generally cited for the AMC, which are the antecedent base-flow index (ABFI), soil moisture index (SMI), and antecedent precipitation index (API) [31]. The ABFI is based on the groundwater runoff, so it is difficult to link it to the soil moisture condition [6]. The SMI is generally applied to the long-term runoff analysis, and thus it is important to consider the evapotranspiration [32–35]. The API considers the precipitation amount during the antecedent 5 to 30 days. The duration can vary, depending on the soil and climate characteristics [25,36]. The SCS–CN method adopted the concept of API, and five days is considered as the proper period for the antecedent precipitation.

In the SCS–CN method, three different AMCs are defined separately for the growing season and the dormant season. The growing season in the US is from June to September, while the dormant season is from October to the following May. For each season, three AMCs are defined by the amount of antecedent five-day precipitation. AMC-I represents the dry condition, thus the direct runoff is minimized. On the other hand, AMC-III represents the wet condition, thus the direct runoff is maximized. AMC-II is the neutral condition in between the wet and dry conditions. These three conditions are considered in the SCS–CN method by controlling the CN. In fact, CN for AMC-III represents the highest CN among those derived by analyzing the sufficient length of rainfall–runoff data. On the other hand, the CN for AMC-I represent the lowest CN. The remaining CNs are then averaged to be CN for the AMC-II [37].

CN for AMC-II (or CN(II)) can be converted into CNs for AMC-I and ANC-III, i.e., AMC(I) and AMC(III), by the following two equations:

$$CN(I) = \frac{4.2\,CN(II)}{10 - 0.058\,CN(II)} \tag{6}$$

$$CN(III) = \frac{23\,CN(II)}{10 + 0.130\,CN(II)} \tag{7}$$

2.2. Estimation of CN by Analyzing the Rainfall–Runoff Data

It is possible to derive CN by analyzing the rainfall–runoff data. The derived CN varies greatly, depending on the rainfall event. Sometimes CN(I) can be estimated higher than CN(II). It is also possible that CN(III) is estimated smaller than CN(II) [38]. This is simply because the possible range of CN(II) is very wide [6,39]. When estimating CN from the observed rainfall–runoff data, some variation may not be avoidable. Based on previous studies [40–43], the CN of a basin is found to decrease as the total precipitation increases. Even though the standard pattern is general [40], another two patterns are also found [40–42]. Various methods for determining the representative CN for a basin have also been proposed by many studies [44–47].

In fact, the estimation of CN for a rainfall event is straightforward if the ratio λ is fixed. Simply the CN can be estimated by solving Equations (4) and (5) with observed P and Q. However, if the ratio λ is unknown, the estimation of CN can be complicated, as these two parameters should be determined by solving the non-linear equation set. Obviously, there is no guarantee of deriving the optimal parameter set. Due to this reason, this study followed the step by step approach. The methodology for estimating CN by analyzing the rainfall–runoff data is summarized as in Figure 1. First, it is important to separate the direct runoff from the total runoff. Fortunately, the main stream of the Hancheon basin is mostly dry, even in the wet summer season. This is a general characteristic of the porous volcanic soil on Jeju Island. The runoff observed at the exit of the Hancheon basin could thus be assumed to be the direct runoff. Second, the initial abstraction is determined by following Woodward et al. [20]. That is, the initial abstraction is determined as the total precipitation amount from the beginning of the rainfall to the occurrence point of the direct runoff. Then the maximum potential retention is estimated by the following equation:

$$S = \frac{(P - I_a)^2}{Q} - (P - I_a) \tag{8}$$

Step 1:	Step 2:	Step 3:	Step 4:
Separation of direct runoff from total runoff	Determination of initial abstraction	Estimation of maximum potential retention	Estimation of CN

Figure 1. The procedure adopted in this study for estimating the parameters of the Soil Conservations Service–Curve Number (SCS–CN) method.

By substituting the initial abstraction estimated in the first step and the observed total precipitation and direct runoff into Equation (8), the maximum potential retention can be calculated. CN can also be calculated using the relation between the maximum potential retention and CN, as in Equation (5).

3. Study Basin and Rainfall Events

3.1. Study Basin

This study considers the Hancheon basin on Jeju Island, Korea (see Figure 2). Here, Hanchcon is the name of the main channel of the Hancheon basin, which originates from Mount Halla, and runs to the seashore in Yongdam-dong. Hancheon is a dry river, which has runoff only during flooding. The upstream part of the basin is mostly forest, the middle stream area is farmland, and the downstream area is urban [4]. Hancheon riverbed is an eroded valley in the basaltic plateau, which has a deep rectangular cross-section, without any embankment. The Hancheon river at the exit of the basin is about 40 m wide and 7 m deep [48]. The area of the Hancheon basin is 32.88 km^2; on the basis of the Second Dongsan Bridge, the channel length is 19.01 km, and the width of the basin is 1.75 km (i.e., the shape factor is 0.09) [13].

Figure 3 shows a picture and cross section of the Second Dongsan Bridge at the exit of the Hancheon basin. There is a surveillance video at the center of the Second Dongsan Bridge, whose image can be obtained from the home page of the Jeju Central Disaster and Safety Countermeasures Headquarters (https://www.jejusi.go.kr/bangjae/realcctv/river.do). An electromagnetic surface flowmeter is also placed at the center of this bridge, which measures both the flow velocity and the water stage every 10 min. The cross-section in Figure 3b was made using the survey data provided in the Master Plan for Local Rivers Management in Jeju-si and the preparation of river facility management [49].

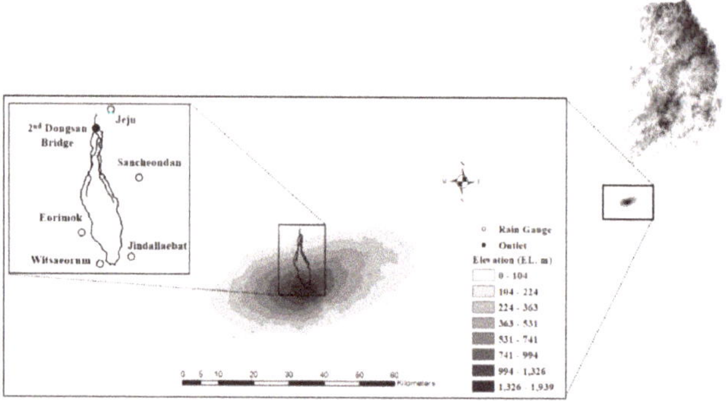

Figure 2. Location of the Hancheon basin in the Jeju Island, Korea (the empty circle represents the rain gauge and the solid circle the stage gauge at the exit of the basin).

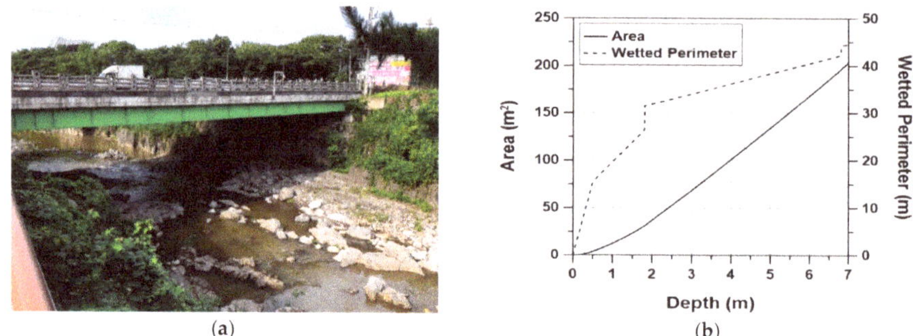

(a)   (b)

Figure 3. Picture of the Second Dongsan Bridge (**a**) and the relation curves of depth-area and depth-wetted perimeter (**b**).

3.2. Data

This study used the 10 min rainfall data obtained from four AWS (Automatic Weather Station) rain gauges around the Hancheon basin (i.e., Jeju (gauge number 184), Sancheondan (329), Eorimok (753), Jindallaebat (870), and Witsaeorum (871)). The areal-average data were then made by applying the Thiessen polygon method. The weighting applied to each rain gauge was Eorimok 48.89%, Sancheondan 22.96%, Jindallaebat 8.64%, Witsaeorum 10.19%, and Jeju 9.32%, respectively.

This study selected a total of ten rainfall events that occurred from 2012 to 2019 (Table 1). As the infiltration capacity of the soil in Jeju Island was known to be very high, only the rainfall events with their total precipitation higher than 100 mm were selected. Several events with significant problems in their runoff records were excluded. Among these selected rainfall events, nine events were caused by typhoons, while the remaining event was convective. In particular, Typhoon Bolaven, which passed the south-western part of Jeju Island in August 2012, recorded a maximum wind velocity of 49.6 m/s at the Seogwipo meteorological station, and a maximum total rainfall of 740.5 mm at the Witsaeorum meteorological station [50]. Based on the areal-average rainfall, the rainfall event #5 recorded the largest total rainfall amount of 691.5 mm, and the rainfall event #3 the highest one-hour rainfall intensity of 76.8 mm/h. The antecedent five-day precipitation for rainfall event #4, which was 564.8 mm, was the highest. Figure 4 compares the rainfall histogram of the rainfall events considered in this study.

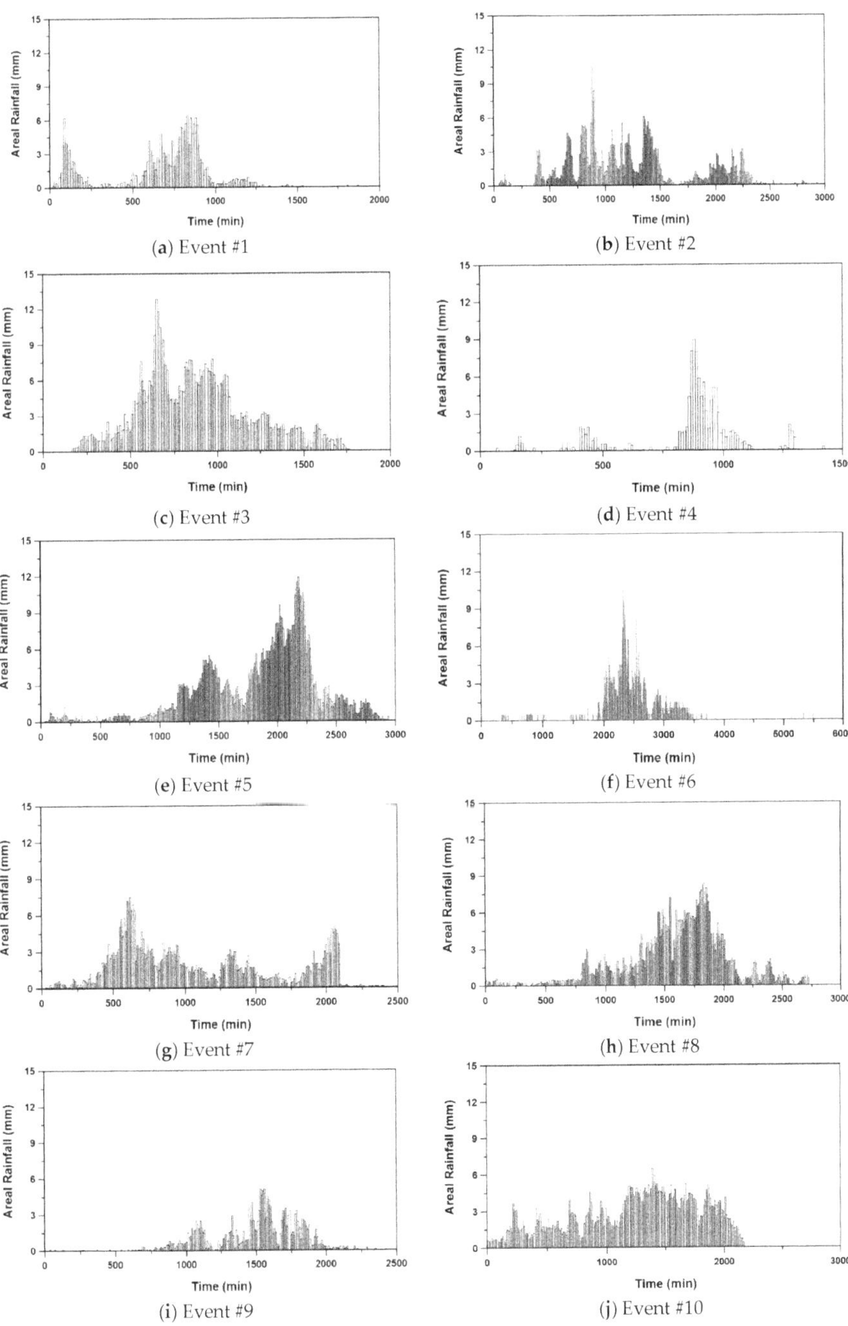

Figure 4. Temporal distribution of the areal-average rainfall over the Hancheon basin for the rainfall events considered in this study.

Table 1. Basic characteristics of the rainfall events considered in this study (these statistics are all for areal average rainfall).

Rainfall Event #	Time (Duration)	Total Rainfall (mm)	Maximum Intensity (mm/h)	Mean Intensity (mm/h)	Comments
1	18 July 2012. 06:00–19 July 2012. 16:00 (34 h)	200.0	38.4	5.9	Typhoon Khanun
2	23 August 2012. 02:00–25 August 2012. 03:00 (49 h)	377.7	63.0	7.7	Convective
3	27 August 2012. 12:00–28 August 2012. 17:00 (29 h)	535.0	76.8	18.4	Typhoon Bolaven
4	29 August 2012. 14:00–30 August 2012. 16:00 (26 h)	113.0	49.2	4.0	Typhoon Tembin
5	15 September 2012. 19:00–17 September 2012. 23:00 (52 h)	691.5	71.4	13.3	Typhoon Sanba
6	31 July 2014. 16:00–04 August 2014. 12:00 (92 h)	400.0	70.8	4.4	Typhoon Nakri
7	22 August 2018. 13:00–24 August 2018. 05:00 (40 h)	398.2	45.0	10.0	Typhoon Soulik
8	4 October 2018. 16:00–6 October 2018. 14:00 (46 h)	480.0	50.4	10.5	Typhoon Kong-rey
9	6 September 2019. 00:00–7 September 2019. 16:00 (40 h)	185.2	30.6	4.6	Typhoon Lingling
10	21 September 2019. 08:00–22 September 2019. 20:00 (36 h)	594.5	39.0	16.4	Typhoon Tapah

The runoff data used were based on the water stage and surface velocity data measured by the electromagnetic surface flowmeter. As the electromagnetic surface flowmeter measures the surface velocity, this should be converted to the cross-sectional average velocity. There have been many studies to determine the conversion factor, or a constant to be used for converting the surface velocity into the cross-sectional average [51–53]. A similar study was also done in Korea for several streams in Jeju Island including the Hancheon [54]. Their result shows that the conversion factor should be 0.48 when the flow depth is lower than 0.5 m, 0.48–0.85 when the flow depth is in between 0.5 and 1.5 m, and 0.85 when the flow depth is higher than 1.5 m. In this study, this conversion factor was used for estimating the cross-sectional average velocity.

Table 2 summarizes the basic characteristics of the runoff data for the ten rainfall events considered. In fact, some portion of the data was missing; in particular, the entire surface velocity data for rainfall event #5 was missing. The missing portions of the data were interpolated by the data observed before and after the missing point. The mean depths of the rainfall events #3 and #5 were rather deep at 1.34 m and 1.43 m, respectively. Rainfall event #8 showed rather high flow velocity of 4.19 m/s.

Table 2. Basic runoff characteristics of the rainfall events in Table 1.

Rainfall Event#	Mean			Maximum		
	Flow Velocity (m/s)	Depth (m)	Discharge (m³/s)	Flow Velocity (m/s)	Depth (m)	Discharge (m³/s)
1	2.39	0.82	98.70	4.90	2.73	187.43
2	2.85	0.92	55.42	7.24	2.72	203.33
3	2.89	1.34	133.79	7.07	4.18	422.71
4	2.23	0.56	53.53	5.87	2.66	134.48
5	n/a	1.43	n/a	n/a	4.52	n/a
6	2.29	0.92	50.11	4.65	2.48	230.28
7	3.22	0.94	89.88	7.23	2.74	289.52
8	4.19	0.88	99.57	7.34	2.53	304.20
9	2.07	0.79	39.51	7.25	1.80	106.88
10	3.28	1.10	87.13	7.31	2.46	188.75

4. Results and Discussions

4.1. Initial Abstraction and CN

The initial abstraction was determined as the rainfall amount from the beginning of the rainfall to the starting point of the direct runoff. However, as most of the time the Hancheon is a dry river, the starting point of the direct runoff coincides with the starting point of the total runoff. The effective rainfall depth was then calculated by dividing the total direct runoff volume by the basin area (i.e., 32.88 km²). CN could then be easily calculated using Equation (5). Figure 5 shows the procedure determining the initial abstraction from the rainfall–runoff data:

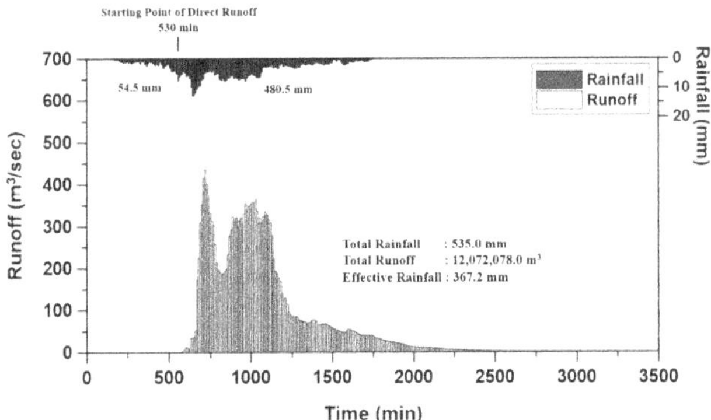

Figure 5. Determination of the initial abstraction in this study (rainfall event #3).

Table 3 summarizes the determined initial abstraction and CN for each rainfall event. CN varies widely from 44.0 to 80.1. As this wide variation was mainly caused by the AMC, the actual variation of CN may be smaller. For example, in the case that CN was estimated to be 80.1 (rainfall event #4), the antecedent five-day precipitation was over 500 mm. On the other hand, in the case of rainfall event #9, CN was estimated to be just 44.0, where the antecedent five-day precipitation was only about 100 mm. In fact, this antecedent five-day precipitation was not a small amount, but could be considered small in a porous basalt region (this issue will be investigated later). Overall, by considering the four rainfall events with more than 100 mm of antecedent five-day rainfall amount, the representative value of CN was assumed to be a little higher than 60.

The maximum potential retention mostly controls the CN, so the variation of the maximum potential retention indicates the variation of CN. However, as the maximum potential retention is dependent on the soil moisture content, it becomes small in the case that CN is high, and vice versa. This relation is also what the SCS-CN method explains, such as in Equation (5). Table 3 also confirms this tendency. For example, rainfall event #4 shows the smallest maximum potential retention of

63.0 mm, but the highest CN of 80.1. Rainfall event #10 shows the highest maximum potential retention of 415.2 mm, but CN was very small at 38.0. The initial abstraction, which is also linearly proportional to the maximum potential retention, could vary, depending on the maximum potential retention.

Table 3. Estimates of initial abstraction, maximum potential retention, and CN along with other basic information.

Rainfall Event#	Antecedent Five-Day Rainfall (mm)	Total Rainfall (mm)	Direct Runoff (mm)	Runoff Ratio	Initial Abstraction (mm)	Maximum Potential Retention (mm)	λ ($I_a=\lambda S$)	CN
1	123.8	200.0	81.0	0.41	39.2	158.4	0.25	61.6
2	88.4	377.7	119.1	0.32	114.9	317.1	0.36	44.5
3	392.2	535.0	367.2	0.69	54.5	148.3	0.37	63.1
4	564.8	113.0	50.5	0.45	25.9	63.0	0.41	80.1
5	10.7	691.5	502.6	0.73	29.8	208.8	0.14	54.9
6	2.2	400.0	182.1	0.46	79.5	243.6	0.33	51.0
7	1.0	398.2	203.7	0.51	72.9	194.2	0.38	56.7
8	0.0	480.0	180.7	0.38	112.6	379.6	0.30	40.1
9	120.5	185.0	47.1	0.25	36.0	323.4	0.11	44.0
10	0.0	594.5	263.2	0.44	107.1	415.2	0.26	38.0

The relation among many variables related to the SCS–CN method can be investigated using the scatterplot (Figure 6). This investigation is also useful to validate the consistency between the estimated variables and the SCS–CN model structure. In this study, the total precipitation, runoff ratio, maximum potential retention, ratio λ, and CN were considered. Among the possible relations, the clearest relation could be found between CN and the maximum potential retention. As mentioned earlier, their negative correlation is what the SCS–CN method adopts. That is, the higher the CN, the smaller the maximum potential retention. Additionally, even though the correlation is rather weak, the negative correlation between the ratio λ and the maximum potential retention was also noticeable. This relation can also be explained by the definition of the initial abstraction. These two negative correlations also indicate the significant positive relation between CN and the ratio λ, which can also be confirmed in Figure 6. Also, as the multiplication of the maximum potential retention and the ratio λ becomes the initial abstraction, no clear relation might be expected between the initial abstraction and CN. The same applies between the total precipitation and CN, and between the runoff ratio and CN. Based on this interpretation, the result summarized in Table 3 can be accepted as valid and consistent with the SCS–CN method.

The basic characteristics of the determined initial abstraction, maximum potential retention, ratio λ, and CN can be easily investigated using the box plot (Figure 7). Table 4 also summarizes the important statistics of the box plot. The box plot shows that the data are not so strongly skewed. There is no outlier detected in all four cases. Symmetricity is strong in the initial abstraction and the maximum potential retention. Even though slightly longer tails are found in the ratio λ and CN, these are mainly due to one or two estimates. For the initial abstraction, the median is 63.7 mm, the first quartile is 36.0 mm, and the third quartile is 107.1 mm. For the maximum potential retention, the median is 226.2 mm, the first quartile is 158.4 mm, and the third quartile is 323.4 mm. The median of ratio λ is 0.32, while the first quartile is 0.25, and the third quartile is 0.37. Finally, the median of CN is 53.0, while the first quartile is 44.0, and the third quartile of 61.6.

Table 4. Quartiles of initial abstraction, maximum potential retention, and CN.

	Initial Abstraction (mm)	Maximum Potential Retention (mm)	λ ($I_a=\lambda S$)	CN
Minimum	25.9	63.0	0.11	38.0
Frist Quartile	36.0	158.4	0.25	44.0
Median	63.7	226.2	0.32	53.0
Third Quartile	107.1	323.4	0.37	61.6
Maximum	114.9	415.2	0.41	80.1

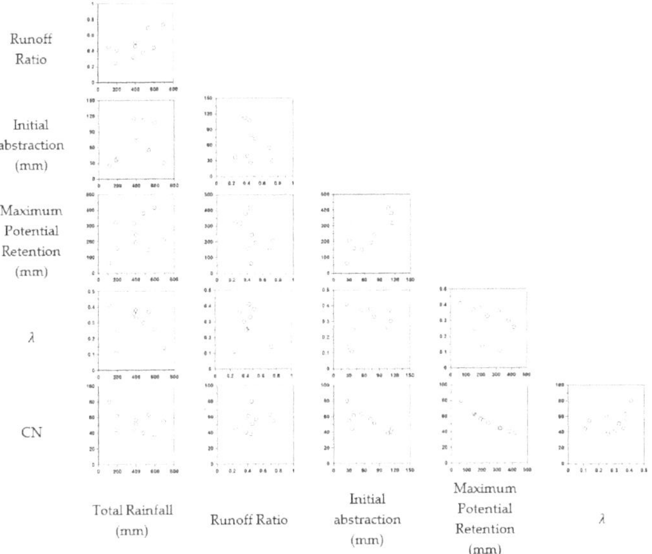

Figure 6. Scatter plots among total rainfall, runoff ratio, initial abstraction, maximum potential retention, λ and curve number (CN).

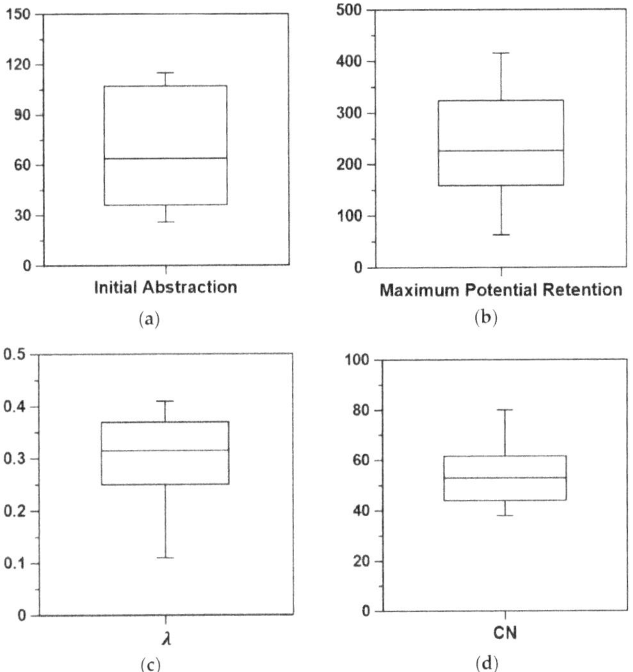

Figure 7. Box-plots of (**a**) initial abstraction, (**b**) maximum potential retention, (**c**) λ, and (**d**) CN.

4.2. Effect of AMC

As mentioned earlier, most variables considered in the SCS–CN method are highly dependent upon the AMC condition. In fact, this investigation of the behavior of these variables with respect to the AMC condition is especially important, as it can provide important tips for determining their representative values. In the previous analysis, it was possible to see their correlation structures and relative behaviors, but it was impossible to determine their representative values.

As the SCS–CN method accepts the concept of API, it should also be important to determine the number of days of antecedent precipitation. The generally accepted period is five days. However, as the Hancheon basin is in a porous basalt region, a slightly shorter period may be more appropriate. This part of the study also aims at this issue, which was investigated as a form of sensitivity analysis. A total of five cases from one to five days were examined in this study. As variables of interest, the maximum potential retention, ratio λ, and CN were considered. In fact, the initial abstraction is related to the AMC condition via all of the three variables to be analyzed.

Table 5 summarizes the antecedent precipitation amount depending on the number of antecedent days to be considered. Basically, the antecedent five-day precipitation amount for rainfall events #5, #6, #7, #8, and #10 was relatively small; in particular, the antecedent precipitation amount for rainfall event #8 and #10 was zero. On the other hand, rainfall events #3 and #4 show a very large antecedent five-day precipitation amount, whose values were 392.2 mm and 564.8 mm, respectively. If considering a slightly shorter period like three days, the antecedent precipitation amount can, in general, become smaller. For example, the antecedent precipitation amount of rainfall events #1, #3, and #9 became much smaller, at 68.0 mm, 47.4 mm, and 49.7 mm, respectively. In particular, the antecedent precipitation amount for rainfall event #3 decreased from 392.2 mm to 47.4 mm.

Table 5. Antecedent n-day rainfall amount for each rainfall event considered in this study.

Rainfall Event#	5-day (mm)	4-day (mm)	3-day (mm)	2-day (mm)	1-day (mm)
1	123.8	103.5	68.0	54.3	44.6
2	88.4	88.0	85.7	64.5	23.7
3	392.2	348.8	47.4	0.2	0.1
4	564.8	534.9	534.8	534.2	12.3
5	10.7	10.7	10.7	7.0	0.2
6	2.2	0.5	0.0	0.0	0.0
7	1.0	1.0	1.0	0.0	0.0
8	0.0	0.0	0.0	0.0	0.0
9	120.5	94.6	49.7	45.6	24.1
10	0.0	0.0	0.0	0.0	0.0

Additionally, the runoff ratio also shows the problem of using the antecedent five-day precipitation amount (see Table 3). The runoff ratios for rainfall events #3, #5, and #7 were estimated to be higher than 0.5, but the others to be less than 0.5. The generally accepted idea is that the runoff ratio is proportional to the antecedent precipitation amount. However, rainfall event #5 showed the highest runoff ratio of 0.73, even though the antecedent five-day precipitation amount was just 10.7 mm. On the other hand, the antecedent five-day precipitation amount of rainfall event #4 was 564.8 mm, but the runoff ratio was 0.45. Basically, the runoff ratio in the Hancheon basin seems to be affected in a complex manner by the rainfall characteristics, including the AMC. For example, the duration of rainfall event #4 was 26 h, but for rainfall event #5, it was 52 h. However, as the total rainfall was 113.0 mm for rainfall event #4 and 691.5 mm for rainfall event #5, the maximum rainfall intensity was (49.2 and 71.4) mm/h, respectively.

Figure 8 shows the behavior of these three variables, i.e., the maximum potential retention, ratio λ, and CN with respect to the antecedent precipitation amount. As can be seen in this figure, the effect of the antecedent precipitation amount becomes clear if the number of days is four or five. In fact, this result is disappointing, as the authors expected a much shorter period, as the Hancheon basin lies in a basalt region. The difference between four or five days was also not clear. The antecedent five-day rainfall amount for the AMC-III condition was found to be higher than 400 mm; on the other hand, that for the AMC-I condition must be less than 100 mm. In fact, this condition for the AMC is far different from that applied to the inland area of mainland Korea. During the growing season in Korea, the antecedent five-day rainfall amount for the AMC-III condition is just 53 mm, while that for the AMC-I condition is 36 mm. Even though the AMC condition for Jeju Island was roughly estimated with a limited number of rainfall events, it was believed to consider adequately the soil characteristics of Jeju Island.

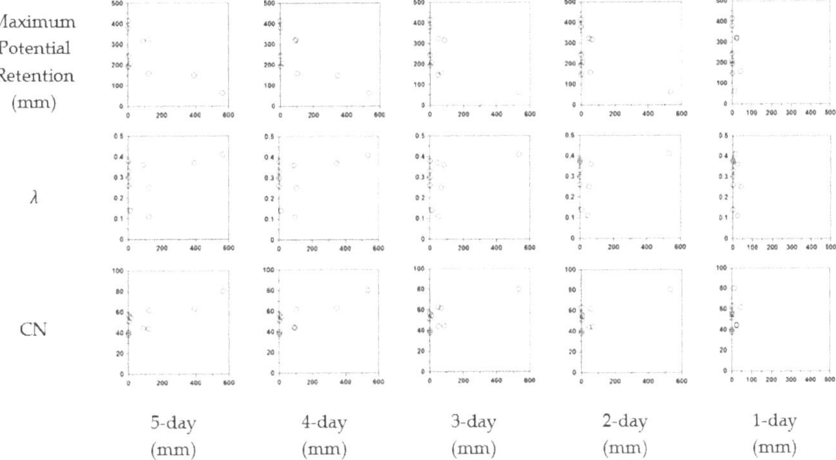

Figure 8. Change of maximum potential retention, λ, and CN depending on the antecedent n-day rainfall amount.

The effect of AMC can be found when considering the antecedent four- or five-day precipitation amount. As mentioned earlier, it was not possible to distinguish these two cases, due to the limited number of rainfall events considered in this study. First of all, an inverse proportional relationship could be found between the AMC and the maximum potential retention. On the other hand, a clear linear proportional relation was found between the AMC and ratio λ, and between the AMC and CN. The maximum potential retention was found to vary from a maximum 300 mm to a minimum 100 mm, which might correspond to the AMC-I and AMC-III conditions, respectively. Under the AMC-II condition, the maximum potential retention might be around 200 mm. The ratio λ was found to vary in the range of 0.2 to 0.4, which might also correspond to the AMC-III and AMC-I conditions, respectively. Their mean value 0.3 might correspond to the AMC-II condition. Lastly, the CN for the AMC-I condition might be 45, and that for the AMC-III condition might be around 80. The CN for the AMC-II condition might be around 65. That is, the CN(I) is near the minimum value of estimated CN values and the CN(III) is near the maximum value. Finally, the CN(II) is the mean value determined for the range of antecedent five-day rainfall amounts between 100 mm and 400 mm. In fact, these CN values under different AMC conditions match well with those given by the SCS–CN method. That is, if the CN value is 65 under the AMC-II condition, the CN(I) and CN(III) become 43.8 and 81.0, respectively [6].

4.3. Classification Rules for Hydrological Soil Group

The hydrological soil group is a key factor in the determination of CN. In general, four different hydrological soil groups are used to represent the infiltration capacity of a soil, where Group A represents the highest infiltration capacity, and Group D the lowest. Groups B and C are in between the two [6]. The hydrological soil group is determined using the reconnaissance soil map or detailed soil map. Various characteristics of the soil provided by these soil maps are used to classify the soil into those four hydrologic soil groups. However, how to consider these soil characteristics to classify the soil or soil series into four different hydrological soil groups is dependent upon the classification rule. That is, the classification rule is not unique, and thus there can be many different classification rules for this same purpose. In Korea, a total of five classification rules are available, among which one is based on the reconnaissance soil map, while the remaining four are based on the detailed soil map.

The classification rule using the reconnaissance soil map is the same as that of the SCS [6]. The reconnaissance soil map provides rather brief information about the soil, among which the main characteristic considered for the determination of the hydrologic soil group is permeability. Permeability is the soil characteristic that controls both the infiltration and drainage capacity.

The four classification rules using the detailed soil map are given by Hu and Jung [55], Jung et al. [56], the Rural Development Administration (RDA) [10], and Lee et al. [57]. First, the classification rule by Hu and Jung [55] considers both the classification rule of the SCS and that in the Soil Survey Manual of the US Department of Agriculture. By considering the soil characteristics, such as infiltration capacity, drainage capacity, permeability, and soil texture in the soil series of the detailed soil map, they could classify these into four hydrologic soil groups.

Jung et al. [56] complemented the classification rule by Hu and Jung [55] and proposed a new system. The most important change of the new system is that it is based on a points system. As a result, the system could exclude subjective judgement, to make the classification result more consistent. Additionally, the authors introduced a new factor of 'impervious layer' to consider the shallow soil depth in Korea. Permeability was also excluded, as it has a similar characteristic to the drainage capacity. As a result, a total of four soil characteristics are evaluated separately by points from 1 to 4. The point of each soil characteristic is then added to derive the total score. If the total score is higher than or equal to 13, the soil series is classified as Group A. On the other hand, if the total point is less than or equal to 7, the soil series is classified as Group D.

RDA [10] additionally complemented the previous studies by considering the in situ experimental results of infiltration and permeability. The in situ experiment was carried out for 58 soil series, which includes 16 forest soils, 17 paddy soils, and 25 farmland and upland soils. Based on these in situ experiment results, the soil series were reclassified to have similar infiltration and permeability characteristics. Additionally, to consider the shallow soil depth in Korea, soil over the impervious layer was classified into Group D, and soil on the limestone into Group C.

Finally, Lee et al. [57] proposed a new method to consider the revised information of the soil series. The revision of the soil series has been carried out by the RDA continuously since 2007 but has not been considered so far in the classification of the hydrologic soil group. In fact, the new classification rule by Lee et al. [57] is basically the same as that by Jung et al. [56] but considered the revision result of the soil series during the last 10 years.

In this study, all five different classification rules were applied to the Hancheon basin to estimate CN. The result shows that the CN using the reconnaissance soil map is 70. Also, the CN as defined by Hu and Jung [55] is 78, that by Jung et al. [56] is 70, that based on the RDA [10] is 77, and that by Lee et al. [57] is 68. As the estimated CN with the observed rainfall–runoff data was around 65, the classification rule by Lee et al. [57] can be assumed to be close to the observed. There has been large variation in the estimated CN depending on the classification rule, so it is just as well that the most recent is closer to the observed.

5. Conclusions

This study investigated three issues in the application of the SCS–CN method to a basin on the volcanic Jeju Island, Korea. The first issue was the possible relation between the initial abstraction, and the maximum potential retention. Determining a proper value for the maximum potential retention was another important issue, which was also closely related to the estimation of CN. The classification rule of the hydrological soil group of soil or soil series was related to this issue. The effects of the AMC on the initial abstraction, maximum potential retention, and CN were the last issues of interest in this study. All of these issues were dealt with based on the analysis of several rainfall events observed in the Hancheon basin, a typical basin on Jeju Island. The results are summarized as follows.

First, estimates of initial abstraction, ratio λ, maximum potential retention, and CN were all found to be consistent with the SCS–CN model structure. That is, CN and the maximum potential retention showed a strong negative correlation, and the ratio λ and the maximum potential retention also showed rather weak negative correlation. On the other hand, a significant positive correlation was found between CN and the ratio λ, which was mainly due to the negative correlations between CN and the maximum potential retention and between the ratio λ and the maximum potential retention. As the multiplication of the maximum potential retention and the ratio λ becomes the initial abstraction, no clear relation could be found between the initial abstraction and CN. The same also applied between the total precipitation and CN, and between the runoff ratio and CN.

Second, the effect of the antecedent precipitation amount was found to be clear if the number of days is four or five. The difference between four and five days was also not clear. The antecedent five-day rainfall amount for the AMC-III condition was found to be higher than 400 mm; on the other hand, that for the AMC-I condition must be less than 100 mm. Even though the AMC condition for Jeju Island was roughly estimated with a limited number of rainfall events, it was believed to consider properly the soil characteristics of Jeju Island. In fact, this condition for the AMC is far different from that applied to the inland area of mainland Korea. During the growing season in Korea, the antecedent five-day rainfall amount for the AMC-III condition is just 53 mm, while that for the AMC-I condition is 36 mm.

Third, an inverse proportional relationship could be found between the AMC and maximum potential retention. On the other hand, a clear linear proportional relation was found between the AMC and ratio λ, and another between the AMC and CN. The maximum potential retention was found to be around 200 mm under the AMC-II condition, and the ratio λ was found to be around 0.3. Also, the CN for the AMC-II condition might be around 65. Additionally, among the five classification rules available in Korea, the classification rule proposed by Lee et al. [57] was found to provide the closest CN to the observed. This result is especially meaningful, as the classification rule proposed by Lee et al. [57] is the most recent. In fact, there have been large overestimations of the CN by up to 78.

In fact, these results derived in the Hancheon basin are all surprising. Especially, the size of the initial abstraction was found to be much higher in the Hancheon basin than for others in the inland area of the Korean Peninsula. The size of the maximum potential retention was also found to be larger, and the ratio λ was also higher, 0.3 rather than 0.2 in the inland area. The role of AMC was also very different. Due to the different soil characteristics of the volcanic Jeju Island, an extremely high antecedent 5-day precipitation amount was found to be valid for defining the AMC-III condition. Interestingly, some of these results were also mentioned in Leta et al. [15]. However, any comment about the AMC could not be found.

Finally, it should be mentioned that the above results were derived by analyzing a limited number of rainfall events. The analysis was also limited to just one basin, even though the authors assumed this basin to be representative of the other river basins on Jeju Island. Despite the practical limitations, the authors are positive about the results derived in this study. With more rainfall events in various river basins in Jeju Island, the results could be consolidated further. This should be future work for the authors of this study.

Author Contributions: C.Y. conceived and designed the idea of this research and wrote the manuscript. M.K. collected the runoff data and rain gauge data and applied the SCS–CN method. All authors have read and agreed to the published version of the manuscript.

Funding: This work was supported by the National Research Foundation of Korea (NRF) grant funded by the Korea government (Ministry of Science and ICT, MSIT) (No. 2020R1A2C200871411).

Conflicts of Interest: The authors declare no conflict of interest.

References

1. Kar, K.K.; Yang, S.K.; Lee, J.H. Assessing unit hydrograph parameters and peak runoff responses from storm rainfall events: A case study in Hancheon Basin of Jeju Island. *JESI* **2014**, *24*, 437–447. [CrossRef]
2. Lau, L.S.; Mink, J.F. *Hydrology of the Hawaiian Islands*; University of Hawaii Press: Honolulu, HI, USA, 2006.
3. Kim, C.G.; Kim, N.W. Evaluation and complementation of observed flow in the Hancheon watershed in Jeju Island using a physically-based watershed model. *J. Korea Water Resour. Assoc.* **2016**, *49*, 951–959.
4. Jeju Special Self-Governing Province. *Comprehensive Investigation of Hydrogeology and Groundwater Resources in Jeju Island (III)*; Jeju Special Self-Governing Province: Jeju, Korea, 2003.
5. Jung, W.; Yang, S. Simulation on runoff of rivers in Jeju Island using SWAT model. *JESI* **2009**, *18*, 1045–1055.
6. Soil Conservation Service (SCS). *National Engineering Handbook—Section 4: Hydrology*; SCS: Washington, DC, USA, 1972.
7. Ministry of Environment (ME). *Design Flood Estimation Tips*; ME: Sejong, Korea, 2019.
8. Jeju Special Self-Governing Province. *Establishment of Jeju-Type River Master Plan and Development of River Facility Management Manual*; Jeju Special Self-Governing Province: Jeju, Korea, 2012.
9. Ha, K.; Park, W.; Moon, D. Estimation of direct runoff variation according to land use changes in Jeju Island. *Econ. Environ. Geol.* **2009**, *42*, 343–356.
10. Rural Development Administration (RDA). *Agro-Environment Research 2006*; RDA: Jeonju, Korea, 2007; pp. 141–176.
11. Jung, W. The Estimation of Parametric Runoff Characteristics and Flood Discharge Based on Riverine in-situ Measurements in Jeju Island. Ph.D. Thesis, Jeju National University, Jeju, Korea, 2013.
12. Sharpley, A.N.; Williams, J.R. *EPIC-Erosion/Productivity Impact Calculator: 1. Model Documentation*; USDA: Washington, DC, USA, 1990.
13. Jung, W.; Yang, S.; Kim, D. Flood discharge to decision of parameters in Han Stream watershed. *JESI* **2014**, *23*, 533–541. [CrossRef]
14. Yang, S.; Kim, M.; Kang, B.; Kim, Y.; Kang, M. Estimation of flood discharge based on observation data considering the hydrological characteristics of the Han Stream basin in Jeju Island. *JESI* **2017**, *26*, 1321–1331. [CrossRef]
15. Leta, O.T.; El-Kadi, A.I.; Dulai, H.; Ghazal, K.A. Assessment of climate change impacts on water balance components of Heeia watershed in Hawaii. *J. Hydrol. Reg. Stud.* **2016**, *8*, 182–197. [CrossRef]
16. Alahmadi, F.; Rahman, N.A.; Yusop, Z. Hydrological modeling of ungauged arid volcanic environments at upper Bathan catchment, Madinah, Saudi Arabia. *J. Teknol.* **2016**, *789*, 9–14.
17. Shehata, M.; Mizunaga, H. Flash flood risk assessment for Kyushu Island, Japan. *Environ. Earth Sci.* **2018**, *77*, 76. [CrossRef]
18. Odongo, V.O.; Mulatu, D.W.; Muthoni, F.K.; van Oel, P.R.; Meins, F.M.; van der Tol, C.; Skidmore, A.K.; Groen, T.A.; Becht, R.; Onyando, J.O.; et al. Coupling socio-economic factors and eco-hydrological processes using a cascade-modeling approach. *J. Hydrol.* **2014**, *518*, 49–59. [CrossRef]
19. Capra, L.; Coviello, V.; Borselli, L.; Márquez-Ramírez, V.H.; Arámbula-Mendoza, R. Hydrological control of large hurricane-induced lahars: Evidence from rainfall-runoff modeling, seismic and video monitoring. *Nat. Hazards Earth Syst. Sci.* **2018**, *18*, 781–794. [CrossRef]
20. Woodward, D.E.; Hawkins, R.H.; Jiang, R.; Hjelmfelt, A.T., Jr.; Van Mullem, J.A.; Quan, Q.D. Runoff curve number method: Examination of the initial abstraction ratio. In Proceedings of the World Water & Environmental Resources Congress 2003, Philadelphia, PA, USA, 23–26 June 2003; pp. 1–10.
21. Baltas, E.A.; Dervos, N.A.; Mimikow, M.A. Technical note: Determination of the SCS initial abstraction ratio in an experimental watershed in Greece. *Hydrol. Earth Syst. Sci.* **2007**, *11*, 1825–1829. [CrossRef]
22. Beck, H.E.; de Jue, R.A.M.; Schellekens, J.; van Dijk, A.I.J.M.; Bruijnzeel, L.A. Improving curve number based strom runoff estimates using soil moisture proxies. *IEEE J. Sel. Top. Appl. Earth Obs. Remote Sens.* **2009**, *2*, 250–259. [CrossRef]
23. Shi, Z.H.; Chen, L.D.; Fang, N.F.; Qin, D.F.; Cai, C.F. Research on the SCS-CN initial abstraction ratio using rainfall-runoff event analysis in the Three Gorges Area, China. *Catena* **2009**, *77*, 1–7. [CrossRef]

24. Yuan, Y.; Nie, W.; McCutcheon, S.C.; Taguas, E.V. Initial abstraction and curve numbers for semiarid watersheds in Southeastern Arizona. *Hydrol. Process.* **2014**, *28*, 774–783. [CrossRef]
25. Ponce, V.M.; Hawkins, R.H. Runoff curve number: Has it reached maturity? *J. Hydrol. Eng.* **1996**, *1*, 11–19. [CrossRef]
26. Babu, P.S.; Mishra, S.K. Improved SCS-CN–inspired model. *J. Hydrol. Eng.* **2012**, *17*, 1164–1172. [CrossRef]
27. Hjelmfelt, A.T.; Kramer, K.A.; Burwell, R.E. Curve numbers as random variables. In Proceedings of the International Symposium on Rainfall-Runoff Modeling; Water Resources Publishing: Littleton, CO, USA, 1982; pp. 365–373.
28. Sorooshian, S.; Duan, Q.; Gupta, V.K. Calibration of rainfall-runoff models: Application of global optimization to the Sacramento Soil Moisture Accounting Model. *Water Resour. Res.* **1993**, *29*, 1185–1194. [CrossRef]
29. Yuan, Y.; Mitchell, J.K.; Hirschi, M.C.; Cooke, R.A. Modified SCS curve number method for predicting subsurface drainage flow. *Trans. ASABE* **2001**, *44*, 1673–1682. [CrossRef]
30. Hawkins, R.H.; Ward, T.J.; Woodward, D.E.; Van Mullem, J.A. *Curve Number Hydrology: State of the Practice*; ASCE: Reston, VA, USA, 2008.
31. Mishra, S.K.; Singh, V.P. *Soil Conservation Service Curve Number (SCS-CN) Methodology*; Springer: Berlin, Germany, 2003.
32. Williams, J.R.; LaSeur, W.V. Water yield model using SCS curve numbers. *J. Hydr. Eng. Div.* **1976**, *102*, 1241–1253.
33. Hawkins, R.H. Runoff curve numbers with varying site moisture. *J. Irrig. Drain. Div.* **1978**, *104*, 389–398.
34. Soni, B.; Mishra, G.C. *Soil Water Accounting Using SCS Hydrological Soil Classification*; National Institute of Hydrology: Roorkee, India, 1985.
35. Mishra, S.K.; Goel, N.K.; Seth, S.M.; Srivastava, D.K. An SCS-CN-based long-term daily flow simulation model for a hilly catchment. In Proceedings of the International Symposium Hydrology of Ungauged Streams in Hilly Regions for Small Hydro Power Development, New Delhi, India, 9–10 March 1998; pp. 9–10.
36. Chen, C.L. An evaluation of the mathematics and physical significance of the soil conservation service curve number procedure for estimating runoff volume. In Proceedings of the International Symposium on Rainfall-Runoff Modeling, Mississippi State, MS, USA, 18–21 May 1982; pp. 387–418.
37. Rallison, R.E.; Miller, N. Past, present, and future SCS runoff procedure. In *Rainfall-Runoff Relationship*; International Symposium on Rainfall-Runoff Modeling: Littleton, CO, USA, 1982; pp. 353–364.
38. Park, C.H.; Yoo, C.; Kim, J.H. Sensitivity analysis of runoff curve number to the rainfall conditions. *J. Korea Soc. Civ. Eng. B.* **2005**, *25*, 501–508.
39. Yoo, C.; Park, C.H.; Kim, J.H. Revised AMC for the application of SCS method: 2. Revised AMC. *J. Korea Water Resour. Assoc.* **2005**, *38*, 963–972. [CrossRef]
40. Hawkins, R.H. Asymptotic determination of runoff curve numbers from data. *J. Irrig. Drain. Eng.* **1993**, *119*, 334–345. [CrossRef]
41. Rezaei-Sadr, H. Influence of coarse soils with high hydraulic conductivity on the applicability of the SCS-CN method. *Hydrol. Sci. J.* **2017**, *62*, 843–848. [CrossRef]
42. Santikari, V.P.; Murdoch, L.C. Including effects of watershed heterogeneity in the curve number method using variable initial abstraction. *Hydrol. Earth Syst. Sci.* **2018**, *22*, 4725–4743. [CrossRef]
43. Soulis, K.X. Estimation of SCS curve number variation following forest fires. *Hydrol. Sci. J.* **2018**, *63*, 1332–1346. [CrossRef]
44. Soulis, K.X.; Valiantzas, J.D. SCS-CN parameter determination using rainfall-runoff data in heterogeneous watersheds-the two-CN system approach. *Hydrol. Earth Syst. Sci.* **2012**, *16*, 1001–1015. [CrossRef]
45. Soulis, K.X.; Valiantzas, J.D. Identification of the SCS-CN parameter spatial distribution using rainfall-runoff data in heterogeneous watersheds. *Water Resour. Manag.* **2013**, *27*, 1737–1749. [CrossRef]
46. Bonta, J.V. Determination of watershed curve number using derived distribution. *J. Irrig. Drain. Eng.* **1997**, *123*, 28–36. [CrossRef]
47. Ajmal, M.; Waseem, M.; Ahn, J.H.; Kim, T.W. Runoff estimation using the NRCS slope-adjusted curve number in mountainous watersheds. *J. Irrig. Drain. Eng.* **2016**, *142*, 04016002. [CrossRef]
48. Yang, S.; Kim, D.; Jung, W. Rainfall-Runoff characteristics in a Jeju stream considering antecedent precipitation. *JESI* **2014**, *23*, 553–560.
49. Jeju Special Self-Governing Province. *Master Plan for Local Rivers Management in Jeju-si and Preparation of River Facility Management*; Jeju Special Self-Governing Province: Jeju, Korea, 2018.

50. Korea Meteorological Administration (KMA). *2012 Typhoon Analysis Report*; KMA: Seoul, Korea, 2013.

51. Costa, J.E.; Spicer, K.R.; Cheng, R.T.; Haeni, F.P.; Melcher, N.B.; Thurman, E.M.; Plant, W.J.; Keller, W.C. Measuring stream discharge by non-contact methods: A proof-of-concept experiment. *Geophys. Res. Lett.* **2000**, *27*, 553–556. [CrossRef]

52. Hauet, A.; Kruger, A.; Krajewski, W.F.; Bradley, A.; Muste, M.; Creutin, J.D.; Wilson, M. Experimental system for real-time discharge estimation using an image-based method. *J. Hydrol. Eng.* **2008**, *13*, 105–110. [CrossRef]

53. Le Coz, J.; Hauet, A.; Pierrefeu, G.; Dramais, G.; Camenen, B. Performance of image-based velocimetry (LSPIV) applied to flash-flood discharge measurements in Mediterranean rivers. *J. Hydrol.* **2010**, *394*, 42–52. [CrossRef]

54. Kim, D.; Yang, S.; Jung, W. Error analysis for electromagnetic surface velocity and discharge measurement in rapid mountain stream flow. *JESI* **2014**, *23*, 543–552.

55. Hu, K.S.; Jung, J.H. Hydrologic classification and its application of the Korean soil. *J. Korean Soc. Agric. Eng.* **1987**, *4*, 48–61.

56. Jung, J.H.; Jang, S.P.; Kim, J.I.; Jung, Y.T.; Hu, K.S.; Park, H. Classification of hydrologic soil group to estimate runoff ratio. *J. Korean Soc. Agric. Eng.* **1995**, *27*, 12–23.

57. Lee, Y.; Kang, M.; Park, C.; Yoo, C. Suggestion of classification rule of hydrological soil groups considering the results of the revision of soil series: A case study on Jeju Island. *J. Korea Water Resour. Assoc.* **2019**, *52*, 35–49.

Publisher's Note: MDPI stays neutral with regard to jurisdictional claims in published maps and institutional affiliations.

Article

Using SCS-CN and Earth Observation for the Comparative Assessment of the Hydrological Effect of Gradual and Abrupt Spatiotemporal Land Cover Changes

Emmanouil Psomiadis [1], Konstantinos X. Soulis [1,*] and Nikolaos Efthimiou [2]

[1] Department of Natural Resources Management and Agricultural Engineering, Agricultural University of Athens, 75 Iera Odos st., 11855 Athens, Greece; mpsomiadis@aua.gr

[2] Faculty of Environmental Sciences, Czech University of Life Sciences Prague, Kamýcká 129, 16500 Praha, Czech Republic; efthimiounik@yahoo.com

* Correspondence: soco@aua.gr; Tel.: +30-2105294070

Received: 21 April 2020; Accepted: 9 May 2020; Published: 13 May 2020

Abstract: In this study a comparative assessment of the impacts of urbanization and of forest fires as well as their combined effect on runoff response is investigated using earth observation and the Soil Conservation Service Curve Number (SCS-CN) direct runoff estimation method in a Mediterranean peri-urban watershed in Attica, Greece. The study area underwent a significant population increase and a rapid increase of urban land uses, especially from the 1980s to the early 2000s. The urbanization process in the studied watershed caused a considerable increase of direct runoff response. A key observation of this study is that the impact of forest fires is much more prominent in rural watersheds than in urbanized watersheds. However, the increments of runoff response are important during the postfire conditions in all cases. Generally, runoff increments due to urbanization seem to be higher than runoff increments due to forest fires affecting the associated hydrological risks. It should also be considered that the effect of urbanization is lasting, and therefore, the possibility of an intense storm to take place is higher than in the case of forest fires that have an abrupt but temporal impact on runoff response. It should be noted though that the combined effect of urbanization and forest fires results in even higher runoff responses. The SCS-CN method, proved to be a valuable tool in this study, allowing the determination of the direct runoff response for each soil, land cover and land management complex in a simple but efficient way. The analysis of the evolution of the urbanization process and the runoff response in the studied watershed may provide a better insight for the design and implementation of flood risk management plans.

Keywords: runoff; SCS-CN; NRCS-CN; earth observation; LUCC; wildfire; urbanization

1. Introduction

Since past decades, impact assessment of land use/cover changes (LUCC) on runoff response attracts a constantly growing attention by the scientific community, due to their immense effect on flood risk increase, soil erosion evolution, hydrological regime and aquatic ecosystems functioning, transfer of pollutants, etc. [1–7]. These impairments classify LUCC as a key factor in global environmental change. In most studies, the terms land use (LU) and land cover (LC) are used collectively (LULC), while in fewer occasions they are interchangeably met. However, they display fundamental differences. Land use refers to the purpose for which the land is used, emphasizing on its functional role on economic activities, without describing the actual cover of the earth's surface [8]. On the other hand, land cover refers to the cover of the earth's surface, i.e., vegetation (natural or agricultural), bare soil or rock, artificial structures and buildings, water, ice, etc. Information on both LU and LC of a given area is a fundamental component of planning and management actions; however, focusing either on LU or LC depends on the purpose of each study. In those related to the LUCC effect on runoff response, like the current one, focus is attracted on LC, considering moreover the land management conditions that depend on LU.

Up till now, numerous studies have underlined the marked effect of LUCC on runoff response [7–16]. Decline of forested areas due to cultivations, urbanization, deforestation, and forest deterioration result in increased direct runoff, total discharge, and peak flows [13–16], which may lead to larger flood risk [7]. Contrary, expanding forest cover may increase evapotranspiration and reduce runoff [9,13]. Extended agricultural areas can increase evapotranspiration and surface flows, leading to a significant decrease in percolation and lateral flow [10,11]. Water abstractions related to agriculture and urban uses may also reduce total discharges [11,12].

Urbanization is one of the most pervasive anthropogenic LU changes, with a range of physical and biochemical impacts on hydrological systems and processes [17]. The global urban population has proliferated from 751 million in 1950 to 4.2 billion in 2018 [18]. The proportion of urban area inhabitants is expected to increase from 55% in 2018 to 68% by 2050 [18]. Thus, understanding the impacts related to urbanization is crucial for the implementation of the "2030 Agenda for Sustainable Development," and the successful management of urban growth. Furthermore, a significant proportion of the recent urban expansion is directed towards the peri-urban environment, i.e., the space at the cities' surroundings that merges into the rural landscape [19].

Several studies have investigated the effects of urbanization-related LUCC on hydrological processes and more importantly on runoff generation [4,17,20–34]. Land development usually entails increased impervious surfaces, considered as the main cause of increased urban surface runoff [20,27], higher peak discharges [28,29], faster runoff response [21], and alteration of the hydrological regimes and the water balance in general [32–34]. Specifically, Hu et al. [24] reported that changes in surface runoff are most strongly correlated with changes on impervious surfaces when compared with the changes of runoff correlated with other types of land use. Furthermore, Miller et al. [21] observed that increase in impervious cover could have a much greater impact on peak flows for rural catchments than for an existing urban area. However, as it is explained by McGrane [23], the heterogeneous urban landscape at contrasting scales is complicating our capacity to quantify infiltration and evapotranspiration dynamics. Additionally, while impervious surfaces exacerbate runoff processes, runoff from pervious areas remains uncertain owing to variable infiltration dynamics.

In addition to the gradual LUCC (e.g., urbanization) effect on the hydrological cycle, several studies have also highlighted the impact of forest fires on the latter, including reduced infiltration and evapotranspiration rates, and increased overland flow [13,35–39]. Forest fires cause abrupt and widespread LUCC, that may have a critical effect on a watershed's hydrological response and associated hydrological risks [35,36,38,40,41]. However, these impacts are (mostly) temporary, and the hydrological processes usually recover in a short period (1 or 2 years) after the event. Under certain conditions, such as harsh meteorological and hydrological regimes, steep relief, highly erodible soils, grazing etc., longer (5–10 years) recovery periods may be required [36,42–44].

One of the most widely used methods for the assessment of LUCC impact on hydrological response is the Soil Conservation Service Curve Number (SCS-CN) method, also known as the US Natural Resources Conservation Service Curve Number (NRCS-CN) method [8,17,20,24,45–47]. SCS-CN is one of the simplest and well-documented conceptual methods for predicting runoff and is used in many hydrologic applications, such as flood design and water balance calculation models. It features easy to obtain and well-documented environmental inputs, and it accounts for many of the factors affecting runoff generation such as soil, land use, and land management, incorporating them in a single CN parameter [45,46,48–53].

Up till now, many research efforts have utilized SCS-CN to quantify the effect of LUCC dynamics on runoff response [8,20,47,54,55]. SCS-CN is routinely used in studies investigating the impact of urbanization on runoff response [17,24,56,57]. Furthermore, the SCS-CN method is generally used in studies estimating the effect of forest fires on the hydrological response of the vegetated areas and the associated hydrological risks [36,40,41,58–64].

Earth Observation (EO) satellites provide high spatial resolution, systematic revisit times and synoptic view of the Earth's surface; hence they are well suited to support several kinds of studies of local, regional, and national interest [65–67]. The multitemporal, high-resolution airborne and satellite data are an accurate and rapid tool to map the spatial distribution of land use/cover and provide a remarkable diachronic historical record for the examination of LULC temporal changes [68–70]. Therefore, they can be utilized as ideal inputs to LULC analyses and hydrological studies [69,71].

Remote sensing techniques can deliver spatial and temporal data of the land surface, appropriate for classification and analysis of watershed areas and runoff models. On the one hand, they can quantify surface factors (such as LULC type, surface roughness, etc.) and their temporal changes, and on the other hand, they can identify hydrologically critical areal phenomena for spatial model output confirmation [67,72]. LUCC in a watershed area with urbanization and deforestation phenomena (mainly due to wildfires), affects the soil's surface characteristics (such as impervious cover and vegetation), that can influence the nature, extent, and speed of surface runoff [70,73]. Various studies have been conducted worldwide utilizing EO information with different methods and datasets, in order to classify and map LULC, urban expansion and wildfire events, and their subsequent impact on the watershed's hydrological characteristics (runoff change, flood frequency, and severity) [72–79].

Starting from the early 1970s, the National Aeronautics and Space Administration (NASA) and the Department of Interior United States Geological Survey (USGS) mission of Landsat satellites, offered the longest continuous space-based record of Earth's surface [80]. Moreover, the European Space Agency (ESA) Copernicus Programme, with the optical satellite Sentinel-2 (S2), made available a new land monitoring and classification mission with 13 spectral bands, having 10 m spatial analysis. The high-resolution multispectral data of Landsat and Sentinel along with very high- resolution aerial imagery, and the CORINE Land Cover (CLC) dataset, constitute a powerful toolbox for LUCC mapping, urban expansion record, wildfire severity, and postfire effects monitoring [67,73,81–86].

To the best of our knowledge, up to now, very few (if any) studies have investigated the combined effect of gradual (e.g., urbanization) and abrupt (e.g., forest fires) LUCC. The purpose of this study is to make a comparative assessment of the impacts of urbanization, causing gradual but lasting LC changes, and of forest fires, causing abrupt but temporary LC changes, as well as their combined effect, on runoff response. To that end, earth observation (to describe the spatiotemporal land cover changes between 1945 and 2018) and the well-known SCS-CN direct runoff estimation method were used. The study is performed at a Mediterranean peri-urban watershed located in the east side of Penteli Mountain, Attica, Greece. The watershed was subjected to several natural and anthropogenic disturbances such as extensive urbanization and forest fires. Furthermore, the existence of an experimental watershed (namely, Lykorema) in the upper part of the study area, which was also affected by a major forest fire, provided the opportunity to use specifically calibrated *CN* values based on real data before and after the fire event effect [36].

2. Materials and Methods

2.1. Study Area

The Rafina watershed is located at the eastern coast of Attica Prefecture, Greece (coordinates: UL 23°45′ E–38°05′ N, LR 24°00′ E–37°56′ N) (Figure 1) approximately 25 km northeast of the Athens city center. It is a peri-urban area in the greater southeast Mesogeia region [87], occupying a surface of 135.32 km². The Hymettus and Penteli mountains are the basin's dominant topographic elements, delimited by shallow catchments of northwest to southeast and with northeast to southwest orientation [88]. Its northern parts are characterized by strong relief (i.e., steep slopes), while the eastern parts and the lowlands towards its outlet to Evoikos Gulf by milder topography. Mean elevation is equal to 228 m, ranging from 0 to 940 m a.s.l. The drainage network forms a quite dense dendritic pattern, developing more complex branches at the eastern and northern regions of the basin. In accordance to the climatic conditions of the broader eastern Attica, local climate is of Mediterranean type (Csa), featuring mild, rainy winters and hot, dry summers with frequent thunderstorms [89]. The bedrock comprises four major lithological units, namely, Neogene formations (Upper Pliocene–Lower Pleistocene), alluvial deposits (Quaternary; Pleistocene), marbles, and limestones [90,91]. Regarding its soils, the upper part of the study area is mostly covered by coarse-textured and very permeable soils, while the lower part is mostly covered by medium- and moderately fine-textured soils and only small parts are covered by fine-textured soils with bad drainage.

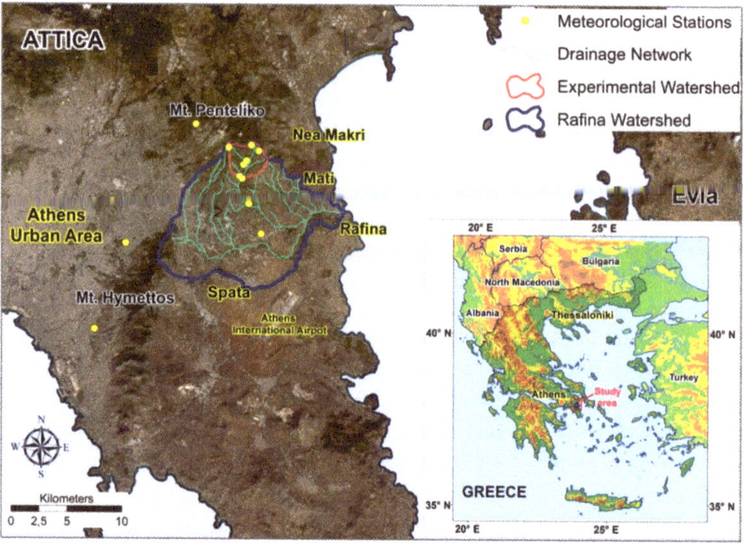

Figure 1. Map of the study area, which is located in Central Greece. The water divides of the Rafina watershed and of the Lykorema experimental watershed as well as the meteorological stations are also presented.

The mean annual precipitation of the study area is approximately equal to 552 mm (2003–2018) ranging from 453 to 665 mm, with the highest depth being met at the northern part of the basin, where the experimental watershed is located, and the lowest at its southern part near the Spata settlement [91]. Snowfall events in the studied area are rare. The dry period extends from May to October. Mean annual temperature fluctuates between 11 and 27 °C.

At the Northern part of the study area; the east side of Penteli Mountain in particular, an experimental watershed is established, operated by the Agricultural University of Athens (AUA), Greece. The latter is equipped with a dense hydrometeorological network, including one hydrometric station at the outlet of the watershed, and five weather stations (Lykorema stream experimental watershed, coordinates: UL 23°53′33″ E–38°04′13″ N, LR 23°56′00″ E–38°02′28″ N) (Figure 1). The watershed is operational from September 2004 up to this day, providing the opportunity to determine the CN values for various land cover types of the study area (especially for natural vegetation and impermeable surface) for the prevalent prefire and postfire conditions, based on real data [35,36,92]. The experimental watershed is characterized by increased permeability with a runoff coefficient for the natural conditions (without the effect of forest fires) of about 0.05 [49]. The Base Flow Index (BFI) for the natural conditions was estimated equal to 0.79. In the period after the fire, the estimated BFI was decreased to 0.54 [35]. The maximum total direct runoff depth observed for the postfire conditions was (54 mm), which is much higher than the maximum total direct runoff depth observed for the natural conditions (7 mm). Finally, the maximum flow rate observed during the postfire period was 81.25 m^3 sec^{-1}, which is considerably higher than the maximum flow rate observed before the fire (6.9 m^3 sec^{-1}) [36].

Details about the characteristics of the experimental watershed can be found in [35,36,49,92,93].

2.2. Background

2.2.1. Urbanization in the Region

Rafina catchment comprises six municipalities of the eastern Attica region, based on the Kallikratis scheme (established in 2011), namely: (a) Rafina—Pikermi (almost entirely, 83.3%), (b) Pallini (almost entirely, 98.4%), (c) Penteli (partly, 56.8%), (d) Spata—Artemida (partly, 31.6%), (e) Paiania (partly, 26.6%), and (f) Agia Paraskevi (partly, 18.3%) [94]. According to the demographic data for the region (1940, 1951, 1981, 1991, 2001, and 2011), a significant increase in the municipalities' population has been recorded at the last 70 years, which in some cases reached values greater than 1000% (Figure 2) [95].

Furthermore, public construction projects of noteworthy scale have been made in the region, especially throughout the last two decades. The most significant of them, involve the development of the new national airport of Athens in Spata, along with an extensive supportive road network, including the Attiki Odos motorway, which connects the Attica Prefecture's center with the southeastern part of Attica, as well as the Rafina port. In addition, various sports facilities were manufactured for the 2004 Olympic Games of Athens, and several accompanying construction projects were established in the area (like shopping malls and leisure facilities) [87].

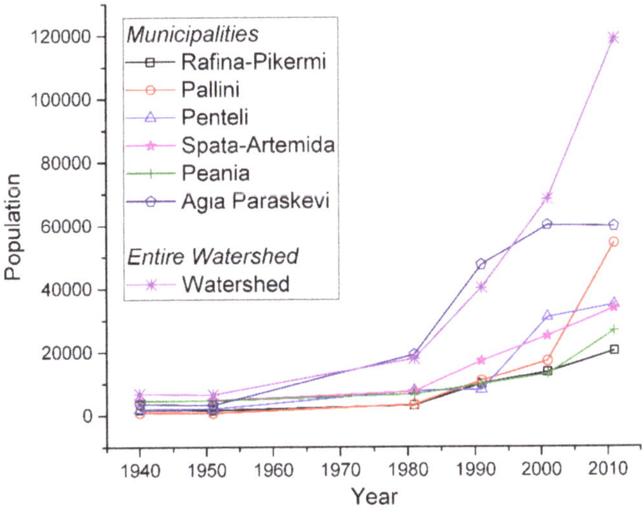

Figure 2. Population growth in the study area according to the demographic data.

2.2.2. Historical Fires in the Region

The basin's northern part is forested with flammable material [59], making it exceptionally prone to fire outbreaks. Over the last decades, a notable number of wildfires has occurred in the broader study area. The fire of July 21, 1995 started at Penteli Mountain, Attica. The event caused the incineration of approximately 251 km^2 of pine forest, and severe damages on almost 105 buildings, yet without loss of human lives [96]. On August 2, 1998, a fire broke out at the northeast suburbs of Athens, reaching the limits of the previously (1995) burned area, and entering "deeper" into the adjacent settlements. Recorded impairments include the reincineration of already regenerated vegetation, forest burn down at the Penteli village outskirts, damages on numerous private and public buildings, and the loss of one human life [97]. The two front fires starting at Rafina, Attica on July 28, 2005 ended up burning 4 km^2

of land and damaging several buildings [97]. The wildfire of August 21, 2009 is considered one of the greatest events, the Attica Prefecture has ever experienced. The fire broke out in the eastern part of the Prefecture, expanding shortly over the entire northeastern Attica, heavily affecting local communities. It burned approximately 850 km^2 of land (pine forest predominantly), damaging 72 houses moreover. The event affected a large part of the experimental watershed of the AUA, providing the opportunity to study the effect of such events on the hydrological response of a basin and to estimate the derivative variation of *CN* values. Finally, on July 23, 2018, one of the most devastating fire events in the history of modern Greece took place. The fire broke out at the forested outskirts of Penteli mountain and descended its way down towards the settlement of Mati, Attica (eastern Attica coast), having affected numerous communities in its path. The extreme weather conditions and specific characteristics of the area (i.e., morphology, microclimate, town configuration, and fuel material) enhanced its magnitude, rapid spread, and overall catastrophic action. The fire caused 101 human casualties, and severe consequences on the local environment (incineration of hundreds of square kilometer area covered by pine forest), personal assets (over 1500 buildings were damaged, and several vehicles were destroyed), infrastructure networks (water supply, telecommunication, and electricity), and regional economic development [91].

2.2.3. Historic Flood Events

The study area drainage network (paired by the public hydraulic works) is considered inadequate to convey flash floods [98]. This fact, in conjunction with the rapid (ongoing) urbanization, is responsible for flood outbreaks, even during precipitation events of less intensity [87]. In this context, four significant flood events of the recent past are presented. The first one occurred on November 2, 1977, classified as the second more important event in the Attica District history, causing severe damages throughout (37 casualties, 1924 houses damaged, 172 cars carried away, and 15% of the road network destroyed) [99]. Although the phenomenon displayed intense manifestation at the study area, no impairments were registered [87]. The event on February 25, 1988 affected the local settlements of Paiania and Marathon, flooding approximately 20 km^2 of land. The flood of March 26, 1998 resulted in the destruction of two bridges [100]. Finally, on February 22, 2013 a flood outbroke caused by a 9-h precipitation event with recorder height of 95.6 and 13.6 mm at the Penteli and Pikermi settlements, respectively [101]. Despite the low depth of the Pikermi event, the induced flood caused several damages (on bridges, houses, and road network) at the regions of Spata and Rafina [87].

The historic floods database of the Greek Ministry of Environment, Energy and Climate Change (MEECC) contains records of additional worth-mentioning flood events in the region [102]. These occurred at the settlements of Pallini (November 8, 1991) and Nea Makri (January 24, 2003), having affected areas very close to the basin's northeastern (November 8,1991, Paiania settlement) and southwestern (January 24, 2003, Kokkino Limanaki settlement) boarders. Unfortunately, no additional information is available regarding the magnitude and derivative effects of the events.

2.3. *Materials*

2.3.1. LULC Earth and Fire Mapping Observation Data

The CLC records and their LULC classification for the years 1990, 2000, 2012, and 2018 were utilized as a base to create the respective maps of the study area [103]. For the delineation of the multi-temporal LULC maps (from 1945 to 2018), several orthophotos datasets were used from 1945 (at the scale of 1:20.000) to 1994, 2007, and 2015 (at the scale of 1:1.000). Landsat-5 (L5, 30 m resolution) and Sentinel-2 (10 m resolution) satellite images of 1986, 1988, 1993, 2000, 2004, and 2018 were also utilized (Table 1).

Moreover, four Landsat 5 Thematic Mapper images were acquired, to record the most severe historical fires that occurred in the area (Table 1). In addition, two S2 images (prefire and postfire) were acquired for the 2018 fire event and the burn scar delineation. The free and open-access

Landsat and Sentinel data were obtained via the EarthExplorer system of the United States Geological Survey (https://earthexplorer.usgs.gov/) and through the European Space Agency open access hub (https://scihub.copernicus.eu/), respectively. The L5 available data were in processing level 1T, having undergone geometrical and standard terrain correction, while the S2 products were in processing level 2A, having been geometrically and atmospherically corrected (Table 1). All satellite images in this study were selected to represent clear atmospheric conditions, without cloud cover. The digital image processing and spatial analysis were accomplished utilizing ENVI (5.5, Harris Geospatial Solutions, USA), ArcGIS (10.5.1, Environmental Systems Research Institute, Redlands, CA, USA), and SNAP (6.0, European Space Agency) software.

Table 1. Earth Observation datasets.

EO System	Instrument	Acquisition Date	Usage
Orthophotos Aerial Imagery	Camera	1945	LULC mapping Scale 1:20.000 (1945) and 1:1.000 (1994, 2007, and 2015)
		1994	
		2007	
		2015	
Landsat 5	Thematic Mapper (TM)	May 22, 1986	LULC mapping (pixel size 30 m)
		August 15, 1988	
		June 10, 1993	
		March 25, 2000	
		September 17, 2000	
		February 01, 2004	
		July 26, 2004	
Landsat 5	Thematic Mapper (TM)	September 13, 1995	Historical Burn scars delineation (pixel size 30 m)
		September 05, 1998	
		September 08, 2005	
		September 03, 2009	
Sentinel-2	Multispectral Instrument (MSI)	July 05, 2018	Burn scar of 2018 wildfire delineation and LULC mapping (pixel size 10 m)
		August 04, 2018	

2.3.2. Soil Data

Due to the lack of a detailed soil map for the entire study area, data from various sources were integrated and interpreted in order to create the required Hydrologic Soil Groups (HSG) map (Figure 3) according to the method documented [104]. The basic information was the soil map of Greece (scale 1:30,000) [105], which, however, covers mostly the cultivated areas. For areas that are not included in the map, various other soil data sources such as the European Soil Database (European Soil Database v2.0, scale 1:1,000,000) [106], topsoil physical properties for Europe based on LUCAS topsoil data [106,107], and information on 14 soil profiles provided by the Greek National Agricultural Research Foundation (NAGREF) were utilized. Regarding the experimental watershed, the detailed soil survey data obtained during the establishment and monitoring of the experimental watershed were utilized [49].

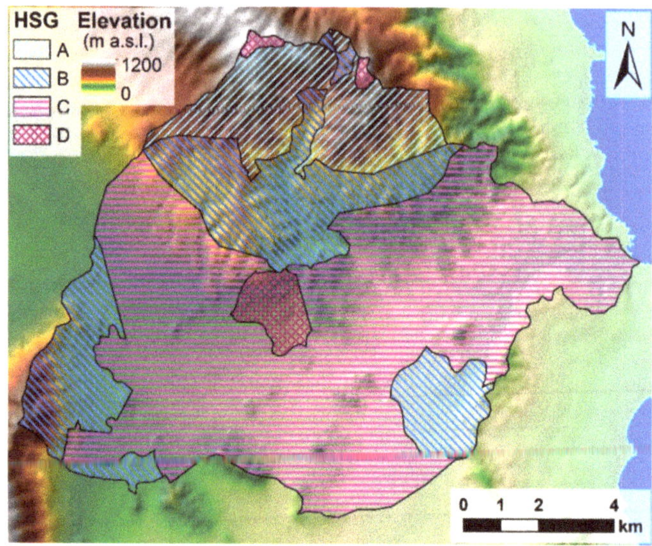

Figure 3. Map of the Hydrologic Soil Groups (HSG) in the studied watershed.

2.3.3. Hydrometeorological Data

Meteorological data were obtained from 13 rain gauge stations located inside or near the studied watershed (Figure 1). Five of them belong to the hydrometeorological network of the experimental watershed, operational since September 2004 [35,36,49]. Eight additional rain gauges, scattered throughout the basin, belong to the METEONET network [101], operational since 2005. In all cases, the data are recorded in a time step of 10 min. Historical precipitation data in a daily time step were also obtained from the Hellenic National Meteorological Service [108].

The hydrological data utilized for the estimation of *CN* values were obtained from the hydrometric station of the experimental watershed. More information about the hydrological data and the corresponding methods are provided in [35,36,109].

2.3.4. Topographic Data

The Hellenic Military Geographical Service (HMGS) provided the required topographic maps ("Kifissia," "Rafina," and "Athinai-Koropi" sheets) in a 1:50,000 scale. The latter were used to extract background information data layers (e.g., geomorphology, rivers, roads, infrastructures, etc.). A detailed DEM, with a spatial resolution of 5 m, 0.5 m geolocation accuracy that was developed based on the 1:5000 scale, and 4 m contour interval topographical maps of HMGS, was also used in this study.

2.3.5. Historical Data of Fires

The diachronic inventory database of forest fires since 1984 (BSM, National Observatory of Athens, http://ocean.space.noa.gr/diachronic_bsm/) and the European Forest Fire Information System (EFFIS, http://effis.jrc.ec.europa.eu/) were used for the cross-checking of significant fire events of the last 40 years in the area.

2.4. Methodology

2.4.1. Wildfires Burn Scar Mapping

The delineation of the wildfire burn scars was divided into two parts: (i) the historical fires that occurred in the area, for which L5 images were utilized and (ii) the recent wildfire of 2018 in Mati [91], for which S2 images were utilized.

Concerning the historical events, L5 images of 1995, 1998, 2005, and 2009 false color composite (FCC) images and the Normalized Difference Vegetation Index were created using the appropriate short-wave (SWIR, band 7), near infrared (NIR, band 5) and red (band 4) spectral bands [FCC: 7,4,2 as RGB and NDVI = (B4 − B3)/(B4 + B3)]. Thereinafter, the burn scars were digitized using visual interpretation procedure.

Regarding the prefire and postfire images of S2, the Normalized Burn Ratio (NBR) index was used to mark burned areas. NBR reveals fire characteristics and evaluates fire impact on vegetation. Healthy vegetation displays very high reflectance in the NIR band, and low reflectance in the shortwave infrared (SWIR) one, and on the other hand, burned areas display the opposite characteristics. High NBR values indicate healthy vegetation in general, whereas low NBR values represent bare ground and recently burned areas [91,110,111]. The NBR formula utilizes the NIR (band 8) and SWIR (band 12) spectral bands (Equation (1)).

$$\mathrm{NBR} = \frac{\mathrm{NIR_{B8}} - \mathrm{SWIR_{B12}}}{\mathrm{NIR_{B8}} + \mathrm{SWIR_{B12}}} \tag{1}$$

Fire frequency is the major controlling factor of vegetation existence and recovery. The information on the historical fire frequency of an area is essential in LUCC and also to explore its impact on the hydrological processes. In sight of the above, a fire frequency map was created by overlaying the polygons of calculated historical burn scars.

2.4.2. Remote Sensing LUCC

LULC was mapped at five different years, based on the CLC records, aerial orthophotos, and satellite images. These data have different spectral and spatial characteristics; therefore, a uniform classification and mapping scale was set prior to the analysis, using the CLC maps. All were resampled to the Greek Grid projection system. CLC is a well-classified continental-scale land cover map, making use of a Minimum Mapping Unit (MMU) of 25 ha. Taking into consideration its limitations (25 ha MMU, higher resolution parcels may be missed, particularly, in urban surrounding regions), a critical "update" was made in entire classes, by additionally introducing a new one named "Urban Forest—UF," in order to distinguish such structures (mainly present at the eastern part of the basin). For that purpose, the very high-resolution aerial orthophotos and satellite images were utilized.

More specifically, the UF class characterizes the Wildland-Urban Interface (WUI) areas, i.e., "the transition zones between cities and wildland, where configurations and other anthropogenic constructions are mixed without clear boundaries with natural wildland or vegetative fuels" (FAO, 2002). According to several researchers, WUI areas constitute a hazardous and associated high-risk ecosystem, in which severe wildfires can provoke vastly disastrous impacts, especially in the Mediterranean region where special attention should be given [91,112,113].

The LULC map of 1945 was created based on the aerial orthophotos of the same year, while the maps of 1990, 2000, 2012, and 2018 by utilizing the corresponding CLC maps, which were updated using mainly aerial orthophotos and in a few cases, supplemental satellite images. The digitization was made by implementing visual photointerpretation technique, with the simultaneous use of additional resources (thematic layers, population censuses, reports, etc.) and information from the burn scars (as described in Section 2.4.1.). As a result, a database of five detailed LULC maps were developed, covering a time period of 73 years (Figure 4).

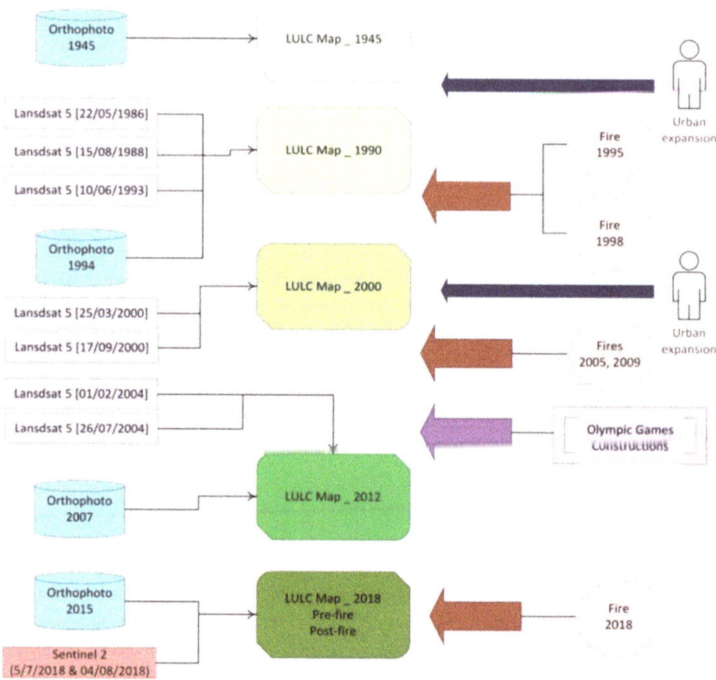

Figure 4. Flowchart of the implemented land use/cover changes (LUCC) methodology.

2.4.3. Runoff Response Estimation

In this study, the Soil Conservation Service Curve Number (SCS-CN) method, which is one of the most widely used and well-documented conceptual methods for the estimation of runoff response of storm events, was used [8,17,20,24,45–47]. This method is a standardized approach in studies investigating the effect of LUCC on runoff response, as well as flood risk assessment and flood design studies [8,17,20,24,36,40,41,47,54–64], because it directly accounts for many of the factors affecting runoff generation such as soil, land use, and land management, incorporating them in a single *CN* parameter [45,46,48–53].

In the SCS-CN method, the direct runoff depth is calculated based on the following equation:

$$Q = \frac{(P-\lambda S)^2}{P+(1-\lambda)S} \quad for\ P > \lambda S$$
$$Q = 0 \quad for\ P \le \lambda S \tag{2}$$

where, P is the total rainfall depth, λ is the initial abstraction ratio, Q is the estimated direct runoff depth, and S is the potential maximum retention depth. When S is expressed in millimeter, it is calculated from the dimensionless curve number (*CN*) parameter using the following relationship:

$$S = \frac{25400}{CN} - 254 \tag{3}$$

The *CN* values corresponding to any given soil, land cover, and land management conditions are typically selected from the various tables included in the method's documentation, and/or the international literature [104]. However, it is always advantageous to estimate *CN* values from available rainfall-runoff data from the studied watershed or nearby ones of similar characteristics.

According to the method's documentation, the determination of the CN values assumes that the initial abstraction rate is a constant value, $\lambda = 0.2$, for CN to be the only unknown parameter of the method. Some studies, having analyzed rainfall-runoff data from several watersheds, suggested that the use of a lower λ value, equal to 0.05, may lead to a better performance [114–116]. In this study, the initial abstraction ratio is set to the typical value ($\lambda = 0.2$) in order to maintain the compatibility with the method documentation.

In order to account for the effect of the soil's antecedent moisture conditions (AMC) on runoff response, the estimated CN values can be adjusted according to the 5-day rainfall preceding the studied event, using the equation described in the method's documentation [106]. The CN values corresponding to the typical conditions (AMC-II) are termed $CNII$, while the ones corresponding to dry (AMC-I) and wet (AMC-III) conditions are termed CNI and $CNIII$, respectively.

Even if AMC was initially considered as the primary source of intrastorm variability, this assumption is of questionable origin, and its use is not directly recommended anymore [117,118]. In the latest version of the documentation, the description of AMC was revised as follows: "*Variability is incorporated by considering the CN as a random variable and the AMC-I and AMC-III categories as bounds of the distribution.*" Other researchers also observed that in many cases, the adaptation of CN value according to AMC category does not improve the runoff prediction [46,48,49,119]. Accordingly, the $CNII$ values were used for the current analysis.

More details on the SCS-CN method can be found in [36,49,53,92,104,109].

As it is explained in detail by Soulis and Valiantzas [109], Grove et al. [120], and Lantz and Hawkins [121], using the SCS-CN method in lumped (i.e., single composite area-weighted CN values for the watershed) and not spatially distributed (i.e., area-weighted runoff estimates from spatially distributed CNs) form can result in significant errors in runoff estimates. The main reason is the nonlinearity of the SCS-CN equation. In this study, the estimation of runoff for each scenario is made in an entirely spatially distributed form using the ArcGIS software package (ESRI, ArcGIS 10.5.1).

To this end, the areal rainfall depth for each examined event was estimated based on the data obtained from the 13 rain gauges (Figure 1), using the elevation regression method. This method was selected considering its simplicity and efficiency, the small size of the studied watershed, and the notably uneven density of the rain gauges in the study area. More specifically, many gauges reside at the upper part of the watershed, where the experimental watershed is located, but only a few at the lower part. Subsequently, the soil/LC complexes' spatial distribution was estimated by overlaying the HSG map (Figure 3) to the respective LC, one for each scenario. The CN values that correspond to each soil–LC complex were estimated using the lookup tables included in the SCS-CN documentation [104] as well as CN values obtained from recorded rainfall-runoff data in the experimental watershed, which is part of the study area.

In most relevant studies, CN values for the various LUCC scenarios are estimated by selecting from the lookup tables, the appropriate LC and management types that correspond to each soil/LC complex [8,17,20,24,47,54–57]. However, in the case of forest fires, CN values for the postfire conditions are typically estimated either by substituting the prefire LC with other LC types that seem more representative of such conditions during the assignment of the CN values from the lookup tables [63,64] or by increasing the CN values of the affected areas [60,122,123] based on general guidelines for postfire CN determination [58,64,123–127].

The main problem is that the CN values corresponding to postfire conditions are not well known. Very few studies have investigated the CN variation following forest fires based on real rainfall-runoff datasets, allowing the comparison between prefire and postfire conditions [125,128], mainly due to the very limited availability of such data. Springer and Hawkins [125] concluded that "*more observations and analyses following fires are needed to support definition of CNs for post-fire response and mitigation efforts.*" Accordingly, this can be a significant source of error. Optimally, CN values should be estimated through the analysis of real hydrological observations comparing prefire and postfire conditions.

In this study, the results of Soulis [36] on the postfire *CN* variation (using detailed prefire and postfire rainfall-runoff datasets in the upper part of the study area (Lykorema experimental watershed)) are utilized in order to increase the accuracy of the yielded results. Specifically, the *CN* values for the soil-LC complexes existing in the experimental watershed for the prefire and postfire conditions were estimated using the "two-CN model" proposed by Soulis and Valiantzas (2012) [109] as it was generalized for heterogeneous watersheds by Soulis and Valiantzas (2013) [92]. Detailed measured rainfall-runoff data for 29 storm events producing significant direct runoff that took place from September 2004 to August 2009, corresponding to prefire conditions, and 60 events that took place from August 2009 to May 2015, corresponding to postfire conditions, were utilized. The approach followed exploits the observed correlation between calculated *CN* values and measured rainfall depths in order to identify the spatial distribution of *CN* values of watersheds taking into account their specific spatial characteristics. Specific details on the *CN* values determination can be found in [36,92,109].

2.4.4. Assessment Steps

In this section, the various steps followed for the comparative assessment of the impacts of urbanization and forest fires on runoff response are presented. The latter comprise:

- Burned area delineation during the study period: The area burned at each forest fire event during the last 40 years (1995, 1998, 2005, 2009, and 2018) was delineated using earth observation data.
- LUCC delineation from 1945 to 2018: To analyze the LUCC, and especially the urbanization and deforestation in the study area, the LC at five points for the years 1945, 1990, 2000, 2012, and 2018 was delineated using earth observation data.
- Design of the investigated scenarios: The scenarios investigated in this study were designed according to the LC and forest fires' time points. The following scenarios were examined: (1) LC-1945, (2) LC-1990, (3) LC-2000, (4) LC-2012, (5) LC-2018, (6) postfire-1995, (7) postfire-1998, (8) postfire-1995+1998, (9) postfire-2005, (10) postfire-2009, (11) postfire-2018, (12) virtual fire-1945, and (13) virtual fire-2018. The postfire scenarios were based on the last available LC before each fire and the burned area of the corresponding event. Regarding the 1995 and 1998 wildfires, the case of combining the burned areas of both events was investigated, since the 1998 fire took place before the complete vegetation recovery. Finally, the two virtual scenarios examine the effect of a virtual forest fire, affecting the area burned by 50% or more of the forest fire events, based on the (a) 1945 LC and (b) 2018 LC. To this end, a fire frequency map was created considering the burned area maps of the five events that occurred during the study period.
- Selection of storm events: Towards selecting a representative set of storm events for the current analysis, all events that took place during the period the Lykorema experimental watershed was monitored were considered. In total, 89 storm events producing significant direct runoff took place from September 2004 to May 2015. The selection of these events was based on the analysis of the hydrograph at the outlet of the experimental watershed. A threshold peak flow rate was identified $(0.25 \text{ m}^3 \text{ s}^{-1})$ based on visual inspection of the latter. The criterion for the separation of the storm events was set to a 3-h interval with rainfall intensity lower than 0.4 mm h^{-1}, considering the concentration time of the watershed. Subsequently, a set of 5 characteristic storm events was selected in order to simplify the analysis. To this end, the events were sorted according to their total rainfall depth, and the greater storm of each year was selected. The resulted rainfall depths were classified into five groups, and the greater event of each group was selected, in order to eliminate those with similar rainfall depths. The resulted areal average total rainfall depths were 160.7, 113.8, 103.9, 91.3, and 72.6 mm. These depths correspond to return periods ranging from 500 to 5 years according to the "Flood Risk Management Plan of the Attica Water District" [129].
- Estimation of runoff: The *CN* spatial distribution for each scenario is calculated, and the runoff response for each event and scenario is estimated using the SCS-CN method, as described in detail

in the previous section. In this manner, the runoff response for each LULC for different rainfall depths is estimated, as well as the increase of runoff after each forest fire.

- Statistical analysis of LUCC and runoff variation: The last step includes the analyses of the LUCC over the studied period (1945–2018), the urbanization and its trends, the affected area by forest fires, and finally the effect of all these factors on runoff response. A comparison between the impact of the gradual (but lasting) LC changes caused by urbanization and of the abrupt (but temporary) LC changes caused by forest fires on runoff response was also performed. Finally, the combined effect of all LUCC is investigated and the trends in runoff response are analyzed.

3. Results

3.1. LUCC-Burned Areas in the Study Area—Deforestation

The statistical analysis of the fire events in the area (performed from 1990 until today) demonstrated the occurrence of one extreme event every 6 years. During the period of interest, the most severe of them, manifested in 1995, 1998, 2005, 2009, and 2018 (one of the deadliest and disastrous wildfires in Greek history). The fires of 1995, 1998, and 2009 affected mainly the northern part of the basin, with the burned area covering 37.1% (45.72 km^2), 27.3% (33.7 km^2), and 27.48% (33.88 km^2) of the basin, respectively. In 1995, the 30.01 km^2 (24.3%) of the burned vegetation was forest and seminatural land cover, while in 1998 and 2009, 22.12 km^2 (19%) and 8.75 km^2 (7.1%) of the same type were affected, respectively. The decrease of the burned forest cover in 1998 and 2009, even if the affected areas were considerably broad, is probably due to the continuous deterioration of vegetation from forest (312, 323, and 324 codes) to sparsely vegetated areas (code 333) type. The events of 2005 and 2018 affected smaller portions of the basin, at its eastern part, covering an area of 7.8 km^2 (6.3%) and 5.9 km^2 (4.8%), respectively (Figure 5). In these two cases, the forest areas burned were small (less than 1.5 km^2); contrary, the affected peri-urban (code 112) and forest urban areas (code 119), in the 2018 wildfire (Table 2), were tremendously large in proportion to the area burned, with dramatic results—especially in the 2018 event (Mati area).

The forest coverage of the area, regarding the coniferous (pine) forest (code 312), increased from 1945 to 1990, mainly due to the abandonment of the cultivated areas (Figure 5). After that year and until 2018, it constantly decreased (25%) because of the forest fires and the population growth. Analogously, the sclerophyllous vegetation and transitional woodland/shrub (code 324) areas were drastically decreased from 33.32 to 6.79 km^2 (79%), mainly converted to sparsely vegetated areas and urban fabric. A small portion of this change involved the transformation of some forest areas to forest urban area at the eastern part of the basin (Table 2 and Figure 6).

From the fire frequency map (Figure 5f), it was noted that approximately 13.68 km^2 (11.1%) and 26.58 km^2 (21.6%) of the total basin area have been burnt two and three times in the time span from 1995 to 2018, respectively, while 1.24 km^2 have been burnt up to five times.

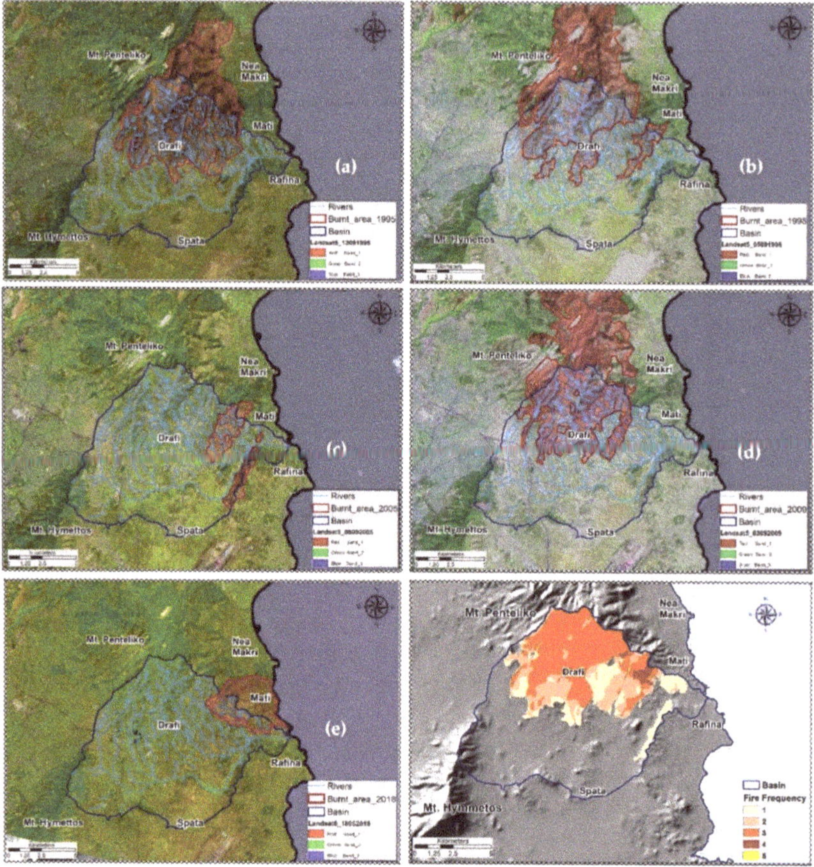

Figure 5. Historical fires in eastern Attica; burned areas delineation using Landsat-5 archive satellite images. (**a**) 1995, (**b**) 1998, (**c**) 2005, (**d**) 2009, and (**e**) 2018. Map (**f**) presents the fire frequency in Rafina watershed (the number of times a surface was impacted by a fire event) during the study period.

3.2. LUCC-Urbanization

Concerning the two principal municipalities occupying 54% of the catchment area, for Rafina-Pikermi, a population increase of 1145% and 645% was estimated from 1940 to 2011 and 1981 to 2011 (from 1769 and from 3134 to 20,266 residents), respectively; while for Pallini, the increase for the periods 1940–2011 and 1981–2011 was 6032% and 1616% (from 902 and from 3367 to 54,415 residents), respectively. The overall population increase in Rafina basin since 1940 was estimated at approximately 1600% (from 7000 to 112,000 residents), while the growth since 1981 was around 620% (Figure 2). This vast urban expansion and the massive increase in construction activity, accompanied by forest vegetation coverage reduction, has drastically altered the landscape and hydrological characteristics of the basin.

The LULC maps analysis revealed a lot of crucial LUCC throughout the time span under consideration. The vast expansion of artificial surfaces (from 2.17 km^2 in 1945 to 45.11 km^2 in 2018), the transformation of forest areas into forest urban areas (3.1 km^2 in 2018), the decrease of the cultivated fields (from 78.70 km^2 in 1945 to 36.20 km^2 in 2018, 46%), and the significant increase of sparsely vegetated areas (from 4.05 km^2 in 1945 to 25.52 km^2 in 2018), as a result of the repeated forest fires that gradually degraded forested areas (Figures 6 and 7a), are the most notable changes in the basin.

The comparison of Figures 2 and 7b leads to the conclusion that there is a direct link between the artificial surfaces and sparsely vegetated areas increment with the population growth. As can be perceived from the fitted sigmoid curve in Figure 7b, the rapid increase of these areas coincides with the outset of rapid population growth (in the early 1980s). It should be noted that no data points are available between 1945 and 1990; however, the sigmoid curve seems to describe better the urbanization process, considering, moreover, the population growth and the information about the development of the studied area. From the analysis of Figure 7b, it is conspicuous that the rate of increase is reducing in the late 2000s; however, this reduction could be partly attributed to the financial crisis that affected Greece at the same period and up to now.

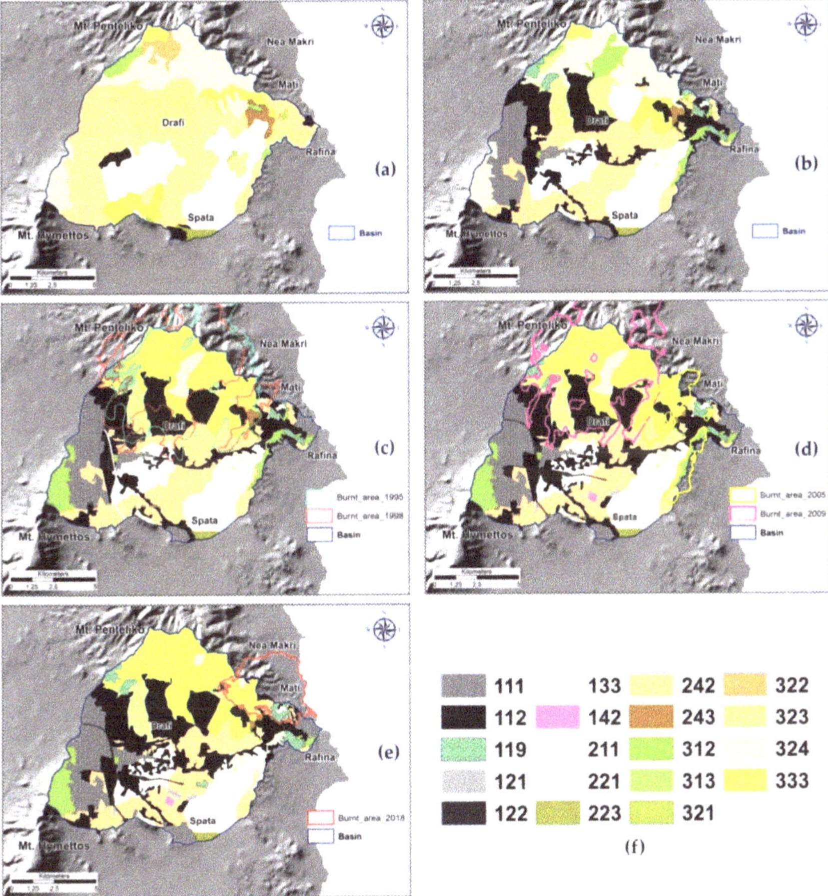

Figure 6. Land use/cover maps of the study area (**a**) 1945, (**b**) 1990, (**c**) 2000, (**d**) 2012, (**e**) 2018 prefire, and (**f**) 2018 postfire; CORINE Land Cover (CLC) codes: 111: Continuous urban fabric/112: Discontinuous urban fabric/119: Urban forest */121: Industrial or commercial units/122: Roads and Rails/133: Construction sites/142: Sport and leisure facilities/211: Non-irrigated arable land/221: Vineyards/223: Olive groves/242: Complex cultivation patterns/243: Land principally occupied by agriculture, with significant areas of natural vegetation/312: Coniferous forest/313: Mixed forest/321: Natural grassland/322: Moors and heathlands/323: Sclerophyllous vegetation/324: Transitional woodland/shrub/333: Sparsely vegetated areas (* the new class 119 developed to describe the specific land use characteristics of the study area).

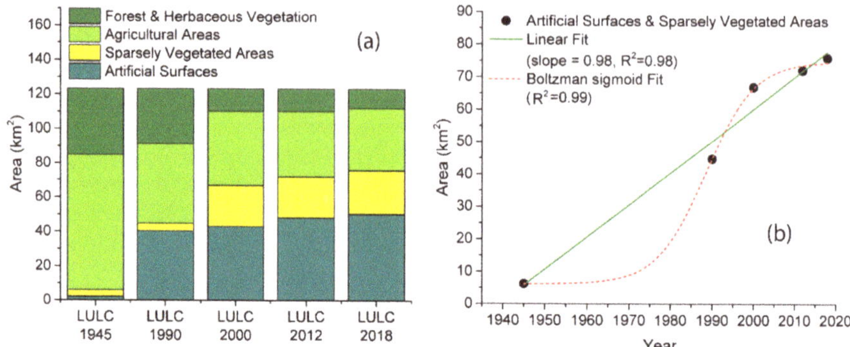

Figure 7. (**a**) Variance of the primary LULC classes coverage and (**b**) evolution of artificial surfaces and sparsely vegetated areas from 1945 to 2018.

Table 2. LULC coverage (Km^2) for the years 1945, 1990, 2000, 2012, and 2018.

Code	Land Use/Cover	1945	1990	2000	2012	2018
111	Continuous urban fabric	0.00	12.42	13.24	13.89	14.87
112	Discontinuous urban fabric	2.17	22.81	27.11	30.03	30.24
119	Urban forest—WUI	0.00	2.22	2.02	2.29	3.10
121	Industrial or commercial units	0.00	0.00	0.00	0.30	0.54
122	Roads and rails	0.00	0.00	0.00	1.32	1.32
133	Construction sites	0.00	2.87	0.58	0.00	0.00
142	Sport and leisure facilities	0.00	0.00	0.00	0.21	0.21
211	Nonirrigated arable land	3.46	0.00	0.00	0.00	0.00
221	Vineyards	24.23	18.07	17.35	15.69	15.16
223	Olive groves	1.01	0.61	0.61	0.61	0.61
242	Complex cultivation patterns	48.39	27.44	24.95	21.55	20.17
243	Land principally occupied by agriculture, with significant areas of natural vegetation	1.61	0.52	0.52	0.25	0.25
312	**Coniferous forest**	**2.50**	**5.70**	**5.69**	**5.79**	**4.28**
313	Mixed forest	0.00	0.00	0.00	0.00	0.22
321	Natural grassland	0.15	0.00	0.00	0.00	0.00
322	Moors and heathlands	2.39	0.00	0.00	0.22	0.00
323	Sclerophyllous vegetation	**12.10**	**6.78**	**3.66**	**3.66**	**3.79**
324	Transitional woodland/shrub	**21.22**	**19.40**	**3.70**	**3.51**	**3.00**
333	Sparsely vegetated areas	4.05	4.45	23.87	23.97	25.52
334	Burnt areas	0.00	0.00	0.00	0.00	0.00
	Total	128.23	128.23	128.23	128.23	128.23

3.3. Runoff Response

The investigation of the impacts of urbanization, of forest fires, and their combined effect on runoff response was based on 13 LULC scenarios and 5 characteristic rainfall events using the SCS-CN direct runoff estimation method. To this end, the *CN* spatial distribution for each scenario was calculated based on the HSG map (Figure 3), on the corresponding LULC map (Figure 6) and, for the case of postfire conditions, on the corresponding burned area map (Figure 5). Characteristic examples of the resulted *CN* maps are presented in Figure 8. Specifically, the *CN* map for the beginning of the studied period that corresponds to preurbanization conditions is presented in Figure 8c and the *CN* map for the ending of the studied period that includes the effect of urbanization is presented in Figure 8a. *CN* values are generally higher in most parts of the watershed, mostly due to the extension of artificial surfaces (lower parts of the watershed) but also due to the degradation of the vegetation cover as a result of the repeated forest fires (north part of the watershed). The spatial average value of *CN* increased from 70 to 80.4 (Table 3). When it comes to the effect of forest fires on *CN* spatial distribution,

CN increment is more prominent in the case of less urbanized conditions (Figure 8). As can be seen in Table 3, where all the results are summarized, the spatial average *CN* value for the same burned area (virtual fire) increases by 5.5 units for 1945 and only by 2.2 units for 2018. It should be noted that (as it is explained in detail in the methodology section) the runoff response was calculated in spatially distributed form and the spatial average *CN* values are only presented as an indication.

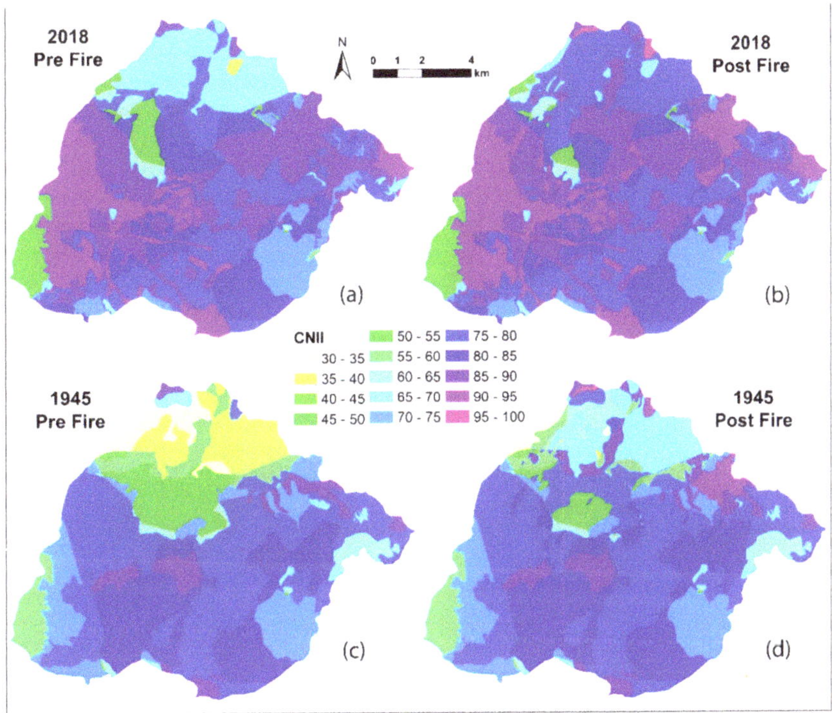

Figure 8. Map of the *CN* values spatial distribution in the study area for four characteristic examples: (**a**) 2018, prefire conditions; (**b**) 2018, postfire conditions; (**c**) 1945, prefire conditions; and (**d**) 1945, postfire conditions.

Table 3. Spatial average *CN* values and runoff response for all the investigated scenarios and storm events.

Fire Event	CNII	Runoff (mm)				
Year	Spatial Average	Event 1 (160.7 mm)	Event 2 (113.8 mm)	Event 3 (103.9 mm)	Event 4 (91.3 mm)	Event 5 (72.6 mm)
1945	70.0	81.0	46.7	40.1	32.0	21.0
1990	74.8	93.7	57.0	49.8	40.8	28.3
1995 postfire	80.6	106.3	65.5	57.3	47.1	32.7
1998 postfire	80.1	105.0	64.3	56.2	46.1	31.9
2005 postfire	79.9	104.9	64.5	56.4	46.4	32.3
2000	79.5	103.6	63.4	55.4	45.4	31.4
2009 postfire	82.4	110.6	68.7	60.2	49.7	34.7
2012	79.9	104.7	64.4	56.3	46.3	32.2
2018	80.4	105.9	65.2	57.1	46.9	32.7
2018 postfire	80.7	106.8	66.0	57.8	47.6	33.2
1945 virtual fire	75.5	92.9	54.5	46.9	37.6	24.8
2018 virtual fire	82.6	111.3	69.4	60.9	50.3	35.4

The calculated runoff response for all examined scenarios and all rainfall events are presented in Table 3. Four characteristic examples of the runoff response's spatial distribution for the larger storm event (Event 1, 160.7 mm) are presented in Figure 9. The higher runoff depths are observed at the higher elevations due to the higher rainfall depths. However, these depths are limited in very narrow areas covered with bare rock or with sparsely vegetated areas combined with very sallow soils. The rest of the upper part of the watershed produces from negligible (Figure 9c) to low (Figure 9a) direct runoff due to the combination of natural vegetation (Figure 6) with permeable soils (Figure 3). The lower parts of the watershed produce from low (Figure 9c) to average (Figure 9a) direct runoff depths. As can be also seen in Figure 9, some parts of the watershed produce the same runoff in all cases. These parts are either agricultural areas or small urban areas that remained unchanged in the study period.

Compering the prefire and postfire runoff response in Figure 9 and Table 3 for 2018 and 1945, it can be observed that the impact of forest fires is much more prominent in 1945. As can be also seen in Table 4, where the runoff increase from the prefire conditions to the postfire conditions is presented for all the forest fire events, the runoff increment was 11.9 mm for the virtual fire in 1945 but only 5.4 mm for the same event in 2018. As it is expected, the maximum runoff depths are observed in the postfire conditions in 2018 for both the actual and the virtual fire event (Table 3 and Figure 9b). Hence, the runoff response for the postfire conditions is already very high due to the urbanization of the watershed.

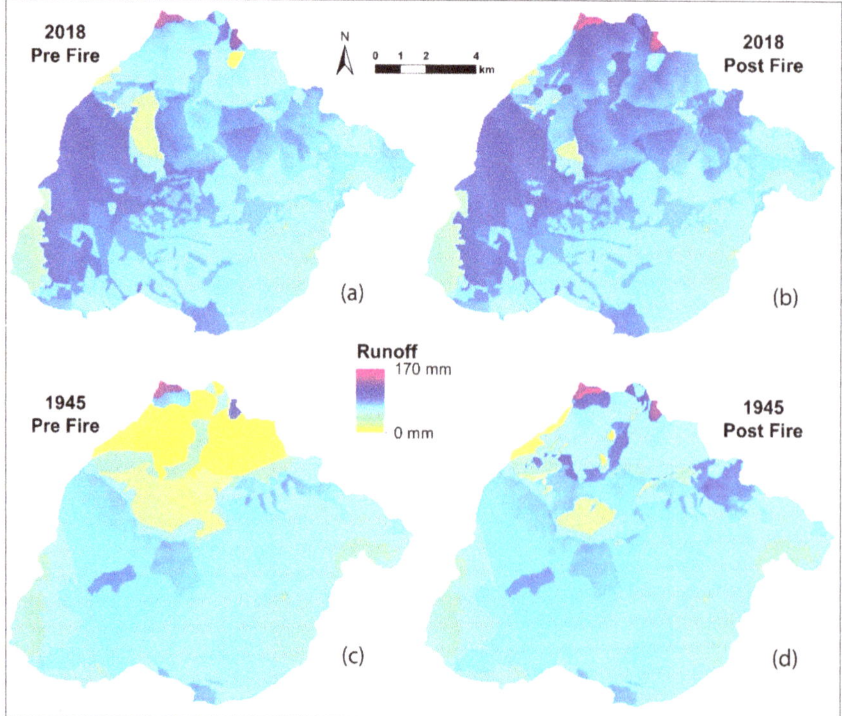

Figure 9. Map of the runoff response spatial distribution in the study area for four characteristic examples: (**a**) 2018, prefire conditions; (**b**) 2018, postfire conditions; (**c**) 1945, prefire conditions; and (**d**) 1945, postfire conditions.

The same observations are also valid for all the storm events. However, as can be seen in Table 4, the percentage changes of runoff response increase as the rainfall depth decreases. For example, the runoff increments due to the 2009 wild fire are 6.4% for Event 1 and 9.4% for Event 5. Similarly, as can be seen in Table 3, the percentage runoff increments due to urbanization from 1945 to 2018 are 31% for Event 1 and 55% for Event 5. Considering also that smaller storm events are much more frequent, this fact may lead to more frequent flood events as the drainage network (paired by the public hydraulic works) in the area is considered inadequate to convey flash floods [98].

An additional important observation in Table 3 is that the runoff increments due to the urbanization of the watershed are much higher than those due to forest fires. Furthermore, the impact of forest fires is smaller in the urbanized watershed as the runoff response is already high before the fire.

Table 4. Runoff increase from the prefire conditions to the postfire conditions for all the forest fire events and for all the studied storm events.

Fire Event	Runoff Increase from the Prefire Conditions (mm (%))				
Year	Event 1 (160.7 mm)	Event 2 (113.8 mm)	Event 3 (103.9 mm)	Event 4 (91.3 mm)	Event 5 (72.6 mm)
Virtual 1945	11.9 (12.8%)	7.8 (14.3%)	6.8 (14.6%)	5.6 (14.9%)	3.8 (15.5%)
1995	12.6 (11.9%)	8.4 (12.9%)	7.5 (13.1%)	6.2 (13.3%)	4.4 (13.5%)
1998	11.3 (10.8%)	7.3 (11.3%)	6.4 (11.4%)	5.2 (11.4%)	3.6 (11.2%)
1995 and 1998	13.6 (12.7%)	9.2 (13.8%)	8.1 (14%)	6.8 (14.3%)	4.9 (14.7%)
2005	1.3 (1.2%)	1.1 (1.7%)	1 (1.8%)	1 (2.1%)	0.8 (2.6%)
2009	7 (6.4%)	5.3 (7.7%)	4.8 (8%)	4.2 (8.5%)	3.3 (9.4%)
2018	2.1 (2%)	1.6 (2.4%)	1.5 (2.6%)	1.3 (2.8%)	1.1 (3.2%)
Virtual 2018	5.4 (4.9%)	4.2 (6%)	3.8 (6.3%)	3.4 (6.8%)	2.7 (7.6%)

The evolution of runoff response as a result of the urbanization evolved from 1945 to 2018 for two characteristic events (events 1 and 5, the larger and the smaller storm depths) are presented in Figure 10a,b. The evolution of runoff response considering the combined effect of urbanization and forest fires for the same events is presented in Figure 10c,d. As can be seen, there is a clear increasing trend in both cases. Interestingly, the slope of the linear fit is almost identical in the case only the effect of urbanization is considered and in the case of the combined effect of urbanization and forest fires. This is an additional indication that urbanization has a dominant impact due to its lasting effect. Furthermore, as can be seen in Figure 10a,b, even if the obtained determination coefficient value is very high, the linear fit does not seem to describe well the evolution of runoff response. Considering also the population growth (Figure 2), and the evolution of the artificial surfaces and sparsely vegetated areas in the studied watershed (Figure 7), the fitting of a sigmoid curve was also attempted. In fact, the determination coefficients indicate an almost perfect fit and the sigmoid curve seems to better describe the urbanization process considering also the population growth as well as information about the developments in the studied area. For example, the rapid runoff response increment during the 1980s and the 1990s is in agreement with the population growth and with the large-scale public constructions that took place in the study area during the same period. Moreover, the stabilization of runoff response since the late 2000s coincides with the financial crisis that affected Greece at the same period. However, it should be noted that more data points are required to obtain safer conclusions. Additionally, any feature projections would be very uncertain because many socioeconomic factors may influence future trends of urbanization and runoff response.

In order to obtain a clearer picture on the impact of urbanization, the relationship between the runoff response and the artificial surfaces area was analyzed (Figure 11a). A strong relationship between the two variables was obtained, however, by the visual inspection of the graph, the increase of artificial surfaces cannot describe entirely the increase of runoff response. Accordingly, the relationship between the runoff response and the sum of artificial surfaces and sparsely vegetated area was also investigated. (Figure 11b). The extension of the artificial surfaces due to the urbanization of the watershed and the increase of sparsely vegetated areas due to the urbanization and the repeated forest fires explains almost entirely the increase of runoff response.

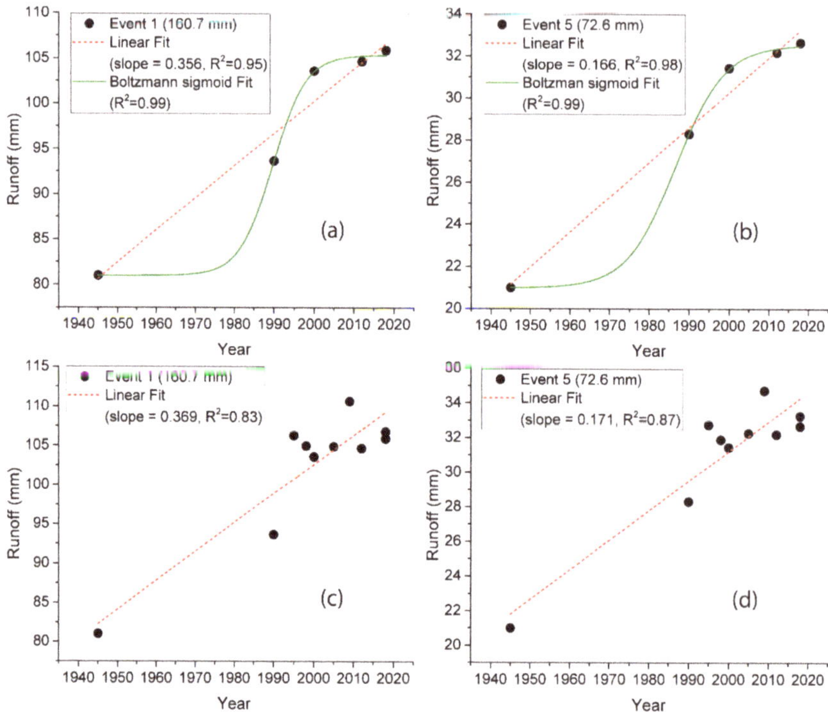

Figure 10. Evolution of the runoff response during the study period considering (**a**,**b**) only the impact of urbanization and (**c**,**d**) the combined effect of urbanization and of forest fires for storm events 1 and 5.

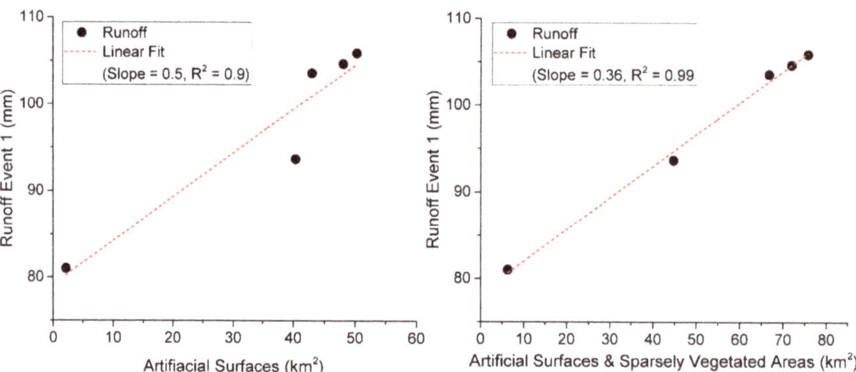

Figure 11. Graph of runoff response against (**left**) artificial surfaces area and (**right**) artificial surfaces and sparsely vegetated areas.

4. Discussion and Conclusions

High and very high-resolution EO data of recent and historical satellite images and aerial photography have excellent potential for delineation of LULC and burn scars to a local scale. Highly accurate LUCC and vegetation features can be obtained from satellite and air photo images of complex landscapes, using various digital-image processing techniques, like photointerpretation and band ratioing. For small-scale catchments, with peri-urban heterogeneous cover affected from continuous wildfires, the high and very high-resolution data offer the advantage of being able to select precisely the information useful at the scale of the hydrological modelling units. In this study, detailed LULC maps, over the previous eight decades, of 1945, 1990, 2000, 2012, and 2018 were created and a thorough LUCC analysis was accomplished. Furthermore, the burned areas of 1995, 1998, 2005, 2009, and 2018 were extracted, and their relation to LUCC was estimated.

The combined use of remote sensing and GIS proved to be an effective tool for LUCC and their interrelationship with urban growth and wildfire events. The systematic and analytical examination of LUCC can be improved to determine the nature, rate, and location of the changes and their possible effects so as to reach an excellent potential for extending environmental management and land-use planning.

The studied watershed is a peri-urban area located approximately 25 km northeast of the Athens city center. This area underwent a significant population increase (Figure 2), a rapid increase of urban LU and a corresponding decrease of forested and agricultural areas (Figure 7a), especially from the 1980s to the early 2000s (Figure 7). Urbanization is considered by far as the fastest type of land use change in Europe, more than 10 times higher than any other LUCC, while peri-urban areas are growing much faster than other urban areas [19]. Urbanization is generally considered as one of the most pervasive anthropogenic LU changes [17]. In this study, it was found that urbanization resulted to a considerable increase of artificial surfaces as well as an increase of sparsely vegetated areas to the detriment of forested areas and of agricultural areas (Figure 7a).

Urban development is usually related with increased impervious surfaces and by extension, increased urban surface runoff [20,27]. The results of the current study indicate that the urbanization process in the studied watershed caused a considerable increase of direct runoff response, especially during the period of fast urban development (Figure 10a,b). It was also observed that percentage runoff increments due to urbanization are much higher for weaker storm events. Taking also into account that weaker storm events are more frequent, a devastating impact on the frequency of flood events is expected as the drainage network (paired by the public hydraulic works) in the area is considered inadequate to convey flash floods [98].

Direct runoff response increase in the studied area is strongly correlated with the extension of artificial surfaces (Figure 11a). However, the extension of sparsely vegetated areas due to the urbanization and the repeated forest fires also has an important contribution (Figure 11b). Hu et al. [24] reported that changes in surface runoff are most strongly correlated with changes in impervious surfaces. Miller et al. [21] observed that increases in impervious cover for rural catchments, as it is also the case of this study, may have a much greater impact on peak flows than for an existing urban area. Furthermore, the combination of increased impervious cover and faster runoff routing via storm drainage network of urbanized areas can lead to a much flashier response and larger peak flows [21]. However, as it is explained by McGrane [23], the heterogeneous urban landscape is complicating the evaluation of runoff response, because, while impervious surfaces exacerbate runoff processes, the variable infiltration dynamics of the remaining pervious areas are still uncertain. For this reason, future study on the refinement of the urban land use typologies used in similar studies may assist to a better representation of the hydrological processes' alteration due to urbanization.

Apart from urbanization, forest fires are the most frequent and widespread disturbance of vegetation, particularly in Mediterranean ecosystems where the areas burned each year are extended and the fire frequency is high [130,131]. A key observation of this study is that the impact of forest fires is much more prominent in rural watersheds than in urbanized watersheds (Figure 9 and Table 3).

However, the increments of runoff response are important during the postfire conditions in all cases. Many other studies also report that forest fires may have a critical effect on vegetated watersheds hydrological response and associated hydrological risks [35,36,38,40,41].

Generally, runoff increments due to urbanization seem to be higher than runoff increments due to forest fires affecting the associated hydrological risks. It should also be considered that the effect of urbanization is lasting and therefore, the possibility of an intense storm to take place is higher than in the case of forest fires that have an abrupt but temporal (generally, 2–5 years [36,42–44]) impact on runoff response. It should be noted though that the combined effect of urbanization and forest fires results in even higher runoff responses, which could be critical for the associated hydrological risks. An additional important observation is that the increase of sparsely vegetated areas has a profound impact as well. Accordingly, except of their direct effect, repeated forest fires may also have a more permanent effect through the gradual degradation of vegetation in forested areas and intensification of soil erosion [91]. Fire frequency seems to be positively related to urban growth, particularly in forest urban interface areas [132].

The SCS-CN method proved to be a valuable tool in this study, allowing the determination of the direct runoff response for soil, land cover, and land management complex in a simple but efficient way, through the estimation of the corresponding *CN* value, based on the sound documentation and the extensive relative literature [45,46,48–53,104]. Up to now, many studies have utilized SCS-CN to quantify the effect of LUCC, e.g., urbanization and forest fires [17,24,36,40,41,56–64], since there is a clear link between the value of its main parameter (*CN*) and LULC facilitating the evaluation of various scenarios.

However, as it is explained in detail by Soulis and Valiantzas [109], the method should be applied in spatially distributed form (i.e., area-weighted runoff estimates from spatially distributed CNs), otherwise, significant errors in runoff estimates can occur. On the positive side, SCS-CN method can be very easily implemented in GIS software tools in such form. An additional advantage is that in this manner, surplus information about the spatial distribution of runoff response is also provided, making it possible to evaluate the impact of specific LUCC at various parts of the watershed.

The obtained results depend on the selected *CN* values for the various soil/LC and management complexes as well as the CN values' adjustments for the postfire conditions. In this study, the existence of the Lykorema experimental watershed in the upper part of the studied area provided the possibility to evaluate the *CN* values used for the various conditions based on real data [36,92,110]. As an example, previous studies on the estimation of the effect of forest fires on the hydrological response, increased *CN* by 5, 10, 15, and 20 units depending on the burn severity [60,121] based on general recommendations that are not based on real hydrological data. However, according to the results presented by Soulis [36] for the Lykorema experimental watershed, the *CN* values during the postfire period in permeable watersheds may increase by 29–36 units depending on the HSG.

This study focused on investigating the impacts of urbanization, forest fires, and their combined effect on runoff response using the SCS-CN direct runoff estimation method. However, the alteration of drainage pathways may also change significantly the storm runoff dynamics. It is reported by Miller et al. [21] that "routing of storm runoff via a storm drainage network can lead to a much flashier response and higher peak flows than would be attributed by increases in impervious cover or change in land use alone." Qi and Liu [3] also reported that studies on impact of LUCC on flood peaks should also consider roughness variations. Accordingly, additional research is needed in order to investigate the effect of gradual and abrupt spatiotemporal LUCC on flood hydrographs and peak flows.

The analysis of the evolution of the urbanization process and the runoff response in the studied watershed may provide a better insight for the design and implementation of flood risk management plans. However, future projections are very uncertain because many socioeconomic factors may influence feature trends of urbanization and runoff response. Hence, it is clear that more policy attention at the regional level and the urban–rural interface is required [19]. Finally, the use of sustainable urban stormwater management solutions such as permeable pavements, green roofs, and other green infrastructure that can attenuate stormwater [133] should also be considered.

Author Contributions: Conceptualization, E.P. and K.X.S.; methodology, E.P. and K.X.S.; software, E.P. and K.X.S.; data analysis, E.P., K.X.S., and N.E.; resources, E.P., K.X.S., and N.E.; writing—original draft preparation, E.P., K.X.S., and N.E.; supervision, E.P.; and field work, K.X.S. All authors have read and agreed to the published version of the manuscript.

Funding: This research received no external funding.

Acknowledgments: The authors acknowledge ESA-Copernicus and USGS-NASA for the availability of Landsat and Sentinel satellite data.

Conflicts of Interest: The authors declare no conflict of interest.

References

1. Snyder, C.D.; Young, J.A. Identification of management thresholds of urban development in support of aquatic biodiversity conservation. *Ecol. Indic.* **2020**, *112*, 106124. [CrossRef]
2. Hanna, D.E.L.; Raudsepp-Hearne, C.; Bennett, E.M. Effects of land use, cover, and protection on stream and riparian ecosystem services and biodiversity. *Conserv. Biol.* **2020**, *34*, 244–255. [CrossRef] [PubMed]
3. Qi, W.; Liu, J. Studies on changes in extreme flood peaks resulting from land-use changes need to consider roughness variations. *Hydrol. Sci. J.* **2019**, *64*, 2015–2024. [CrossRef]
4. Wang, H.; Chen, Y. Identifying key hydrological processes in highly urbanized watersheds for flood forecasting with a distributed hydrological model. *Water* **2019**, *11*, 1641. [CrossRef]
5. Khan, M.; Sharma, A.; Goyal, M.K. Assessment of future water provisioning and sediment load under climate and LULC change scenarios in a peninsular river basin, India. *Hydrol. Sci. J.* **2019**, *64*, 405–419. [CrossRef]
6. Zhang, X.; Lin, P.; Chen, H.; Yan, R.; Zhang, J.; Yu, Y.; Liu, E.; Yang, Y.; Zhao, W.; Du, L.; et al. Understanding land use and cover change impacts on run-off and sediment load at flood events on the Loess Plateau, China. *Hydrol. Process.* **2018**, *32*, 576–589. [CrossRef]
7. Yang, X.; Chen, H.; Wang, Y.; Xu, C. Evaluation of the effect of land use/cover change on flood characteristics using an integrated approach coupling land and flood analysis. *Hydrol. Res.* **2016**, *47*, 1161–1171. [CrossRef]
8. Vojtek, M., Vojteková, J. Land use change and its impact on surface runoff from small basins: A case of Radiša basin. *Folia Geogr.* **2019**, *61*, 104–125.
9. Xu, F.; Jia, Y.; Peng, H.; Niu, C.; Liu, J.; Hao, C.; Huang, G. Vertical zonality of the water cycle and the impact of land-use change on runoff in the Qingshui River Basin of Wutai Mountain, China. *Hydrol. Sci. J.* **2019**, *64*, 2080–2092. [CrossRef]
10. Martínez-Retureta, R.; Aguayo, M.; Stehr, A.; Sauvage, S.; Echeverría, C.; Sánchez-Pérez, J. Effect of land use/cover change on the hydrological response of a southern center basin of Chile. *Water* **2020**, *12*, 302. [CrossRef]
11. Kang, Y.; Gao, J.; Shao, H.; Zhang, Y. Quantitative analysis of hydrological responses to climate variability and land-use change in the hilly-gully region of the Loess Plateau, China. *Water* **2020**, *12*, 82. [CrossRef]
12. Tuladhar, D.; Dewan, A.; Kuhn, M.; Corner, R.J. The influence of rainfall and land use/land cover changes on river discharge variability in the mountainous catchment of the Bagmati river. *Water* **2019**, *11*, 2444. [CrossRef]
13. Santos, R.M.B.; Fernandes, L.F.S.; Cortes, R.M.V.; Pacheco, F.A.L. Hydrologic impacts of land use changes in the Sabor river basin: A historical view and future perspectives. *Water* **2019**, *11*, 1646. [CrossRef]
14. Wang, Q.; Xu, Y.; Xu, Y.; Wu, L.; Wang, Y.; Han, L. Spatial hydrological responses to land use and land cover changes in a typical catchment of the Yangtze river delta region. *Catena* **2018**, *170*, 305–315. [CrossRef]
15. Truong, N.C.Q.; Nguyen, H.Q.; Kondoh, A. Land Use and Land Cover Changes and their effect on the flow regime in the Upstream Dong Nai River Basin, Vietnam. *Water* **2018**, *10*, 1206. [CrossRef]
16. Robinet, J.; Minella, J.P.G.; de Barros, C.A.P.; Schlesner, A.; Lücke, A.; Ameijeiras-Mariño, Y.; Opfergelt, S.; Vanderborght, J.; Govers, G. Impacts of forest conversion and agriculture practices on water pathways in southern Brazil. *Hydrol. Process.* **2018**, *32*, 2304–2317. [CrossRef]

17. Li, C.; Liu, M.; Hu, Y.; Shi, T.; Zong, M.; Walter, M.T. Assessing the impact of urbanization on direct runoff using improved composite CN method in a large urban area. *Int. J. Environ. Res. Public Health* **2018**, *15*, 775. [CrossRef]

18. United Nations. 2018 Revision of World Urbanization Prospects. United Nations Department of Economic and Social Affairs. 2018. Available online: https://www.un.org/development/desa/publications/2018-revision-of-world-urbanization-prospects.html (accessed on 2 March 2020).

19. Piorr, A.; Ravetz, J.; Tosics, I. (Eds.) Peri-urbanisation in Europe: Towards European Policies to Sustain Urban-Rural Futures. 2011. Available online: http://www.openspace.eca.ed.ac.uk/wp-content/uploads/2015/12/Peri_Urbanisation_in_Europe_printversion.pdf (accessed on 1 March 2020).

20. Li, F.; Chen, J.; Liu, Y.; Xu, P.; Sun, H.; Engel, B.A.; Wang, S. Assessment of the impacts of land use/cover change and rainfall change on surface runoff in China. *Sustainability* **2019**, *11*, 3535. [CrossRef]

21. Miller, J.D.; Kim, H.; Kjeldsen, T.R.; Packman, J.; Grebby, S.; Dearden, R. Assessing the impact of urbanization on storm runoff in a peri-urban catchment using historical change in impervious cover. *J. Hydrol.* **2014**, *515*, 59–70. [CrossRef]

22. Lin, Z.; Xu, Y.; Dai, X.; Wang, Q.; Gao, B.; Xiang, J.; Yuan, J. Changes in the plain river system and its hydrological characteristics under urbanization–case study of Suzhou city, China. *Hydrol. Sci. J.* **2019**, *64*, 2068–2079. [CrossRef]

23. Mc Grane, S.J. Impacts of urbanization on hydrological and water quality dynamics, and urban water management: A review. *Hydrol. Sci. J.* **2016**, *61*, 2295–2311. [CrossRef]

24. Hu, S.; Fan, Y.; Zhang, T. Assessing the effect of land use change on surface runoff in a rapidly urbanized city: A case study of the central area of Beijing. *Land* **2020**, *9*, 17. [CrossRef]

25. Yang, L.; Feng, Q.; Yin, Z.; Wen, X.; Si, J.; Li, C.; Deo, R. Identifying separate impacts of climate and land use/cover change on hydrological processes in upper stream of Heihe River, Northwest China. *Hydrol. Process.* **2017**, *31*, 1100–1112. [CrossRef]

26. Berihun, M.; Tsunekawa, A.; Haregeweyn, N.; Meshesha, D.; Adgo, E.; Tsubo, M.; Masunaga, T.; Fenta, A.; Sultan, D.; Sebhat, M.; et al. Hydrological responses to land use/land cover change and climate variability in contrasting agro-ecological environments of the Upper Blue Nile basin, Ethiopia. *Sci. Total Environ.* **2019**, *689*, 347–365. [CrossRef] [PubMed]

27. Wei, L.; Hubbart, J.A.; Zhou, H. Variable streamflow contributions in nested subwatersheds of a US Midwestern urban watershed. *Water Resour. Manag.* **2017**, *32*, 1–16. [CrossRef]

28. Rose, S.; Peters, N.E. Effects of urbanization on streamflow in the Atlanta area (Georgia, USA): A comparative hydrological approach. *Hydrol. Process.* **2010**, *15*, 1441–1457. [CrossRef]

29. Huang, H.; Cheng, S.; Wen, J.; Lee, J. Effect of growing watershed imperviousness on hydrograph parameters and peak discharge. *Hydrol. Process.* **2010**, *22*, 2075–2085. [CrossRef]

30. Yao, L.; Wei, W.; Yu, Y.; Xiao, J.; Chen, L. Rainfall-runoff risk characteristics of urban function zones in Beijing using the SCS-CN model. *J. Geogr. Sci.* **2018**, *28*, 656–668. [CrossRef]

31. Guan, M.; Sillanpää, N.; Koivusalo, H. Storm runoff response to rainfall pattern, magnitude and urbanization in a developing urban catchment. *Hydrol. Process.* **2016**, *30*, 543–557. [CrossRef]

32. Zhang, Y.; Xia, J.; Yu, J.; Randall, M.; Zhang, Y.; Zhao, T.; Pan, X.; Zhai, X.; Shao, Q. Simulation and assessment of urbanization impacts on runoff metrics: Insights from land use changes. *J. Hydrol.* **2018**, *560*, 247–258. [CrossRef]

33. Braud, I.; Breil, P.; Thollet, F.; Lagouy, M.; Branger, F.; Jacqueminet, C.; Kermadi, S.; Michel, K. Evidence of the impact of urbanization on the hydrological regime of a medium-sized periurban catchment in France. *J. Hydrol.* **2013**, *485*, 5–23. [CrossRef]

34. Michael, O.D.; Sandra, C.; Anne, J.; Alex, M.; Sara, M. Urbanization effects on watershed hydrology and in-stream processes in the Southern United States. *Water* **2010**, *2*, 605–648.

35. Soulis, K.X.; Dercas, N.; Valiantzas, J.D. Wildfires impact on hydrological response—The case of Lykorrema experimental watershed. *Glob. NEST J.* **2012**, *14*, 303–310.

36. Soulis, K.X. Estimation of SCS Curve Number variation following forest fires. *Hydrol. Sci. J.* **2018**, *63*, 1332–1346. [CrossRef]

37. Ponomarev, E.I.; Ponomareva, T.V.; Prokushkin, A.S. Intraseasonal dynamics of river discharge and burned forest areas in Siberia. *Water* **2019**, *11*, 1146. [CrossRef]

38. Folton, N.; Andréassian, V.; Duperray, R. Hydrological impact of forest-fire from paired-catchment and rainfall–runoff modelling perspectives. *Hydrol. Sci. J.* **2015**, *60*, 1213–1224. [CrossRef]

39. Zhou, Y.; Zhang, Y.; Vaze, J.; Lane, P.; Xu, S. Impact of bushfire and climate variability on streamflow from forested catchments in southeast Australia. *Hydrol. Sci. J.* **2015**, *60*, 1340–1360. [CrossRef]

40. Candela, A.; Aronica, G.; Santoro, M. Effects of forest fires on flood frequency curves in a Mediterranean catchment. *Hydrol. Sci. J.* **2005**, *50*, 193–206. [CrossRef]

41. Nalbantis, I.; Lymperopoulos, S. Assessment of flood frequency after forest fires in small ungauged basins based on uncertain measurements. *Hydrol. Sci. J.* **2012**, *57*, 52–72. [CrossRef]

42. Cerda, A. Changes in overland flow and infiltration after a rangeland fire in Mediterranean scrubland. *Hydrol. Process.* **1998**, *12*, 1031–1042. [CrossRef]

43. Mayor, A.G.; Bautista, S.; Llovet, J.; Bellot, J. Post-fire hydrological and erosional responses of a Mediterranean landscape: Seven years of catchment-scale dynamics. *Catena* **2007**, *71*, 68–75. [CrossRef]

44. Li, X.; Xie, F.; Wang, X.; Kong, F. Human intervened post-fire forest restoration in the Northern Great Hing'an Mountains: A review. *Landsc. Ecol. Eng.* **2006**, *2*, 129–137. [CrossRef]

45. Ling, L.; Yusop, Z.; Yap, W.-S.; Tan, W.L.; Chow, M.F.; Ling, J.L. A calibrated, watershed-specific SCS-CN method: Application to Wangjiaqiao Watershed in the Three Gorges Area, China. *Water* **2020**, *12*, 60. [CrossRef]

46. Krajewski, A.; Sikorska-Senoner, A.E.; Hejduk, A.; Hejduk, L. Variability of the Initial Abstraction Ratio in an urban and an agroforested catchment. *Water* **2020**, *12*, 415. [CrossRef]

47. Sultan, D.; Tsunekawa, A.; Haregeweyn, N.; Adgo, E.; Tsubo, M.; Meshesha, D.T.; Masunaga, T.; Aklog, D.; Ebabu, K. Analyzing the runoff response to soil and water conservation measures in a tropical humid Ethiopian highland. *Phys. Geogr.* **2017**, *38*, 423–447. [CrossRef]

48. Steenhuis, T.S.; Winchell, M.; Rossing, J.; Zollweg, J.A.; Walter, M.F. SCS runoff equation revisited for variable-source runoff areas. *J. Irrig. Drain. Eng.* **1995**, *121*, 234–238. [CrossRef]

49. Soulis, K.X.; Valiantzas, J.D.; Dercas, N.; Londra, P.A. Analysis of the runoff generation mechanism for the investigation of the SCS-CN method applicability to a partial area experimental watershed. *Hydrol. Earth Syst. Sci.* **2009**, *13*, 605–615. [CrossRef]

50. Van Dijk, A.I.J.M. Selection of an appropriately simple storm runoff model. *Hydrol. Earth Syst. Sci.* **2010**, *14*, 447–458. [CrossRef]

51. Santra Mitra, S.; Wright, J.; Santra, A.; Ghosh, A.R. An integrated water balance model for assessing water scarcity in a data-sparse interfluve in eastern India. *Hydrol. Sci. J.* **2015**, *60*, 1813–1827. [CrossRef]

52. Rezaei-Sadr, H. Influence of coarse soils with high hydraulic conductivity on the applicability of the SCS-CN method. *Hydrol. Sci. J.* **2017**, *62*, 843–848. [CrossRef]

53. Verma, S.; Verma, R.K.; Mishra, S.K.; Singh, A.; Jayaraj, G.K. A revisit of NRCS-CN inspired models coupled with RS and GIS for runoff estimation. *Hydrol. Sci. J.* **2017**, *62*, 1891–1930. [CrossRef]

54. Al-Juaidi, A.E.M. A simplified GIS-based SCS-CN method for the assessment of land-use change on runoff. *Arab. J. Geosci.* **2018**, *11*, 269. [CrossRef]

55. Rizeei, H.M.; Pradhan, B.; Saharkhiz, M.A. Surface runoff prediction regarding LULC and climate dynamics using coupled LTM, optimized ARIMA, and GIS-based SCS-CN models in tropical region. *Arab. J. Geosci.* **2018**, *11*, 53. [CrossRef]

56. Myronidis, D.; Ioannou, K. Forecasting the urban expansion effects on the design storm hydrograph and sediment yield using artificial neural networks. *Water* **2019**, *11*, 31. [CrossRef]

57. Hameed, H.M. Estimating the effect of urban growth on annual runoff volume using GIS in the Erbil sub-basin of the Kurdistan region of Iraq. *Hydrology* **2017**, *4*, 12. [CrossRef]

58. Goodrich, D.C.; Evan Canfield, H.; Shea Burns, I.; Semmens, D.J.; Miller, S.N.; Hernandez, M.; Levick, L.R.; Guertin, D.P.; Kepner, W.G. Rapid post-fire hydrologic watershed assessment using the AGWA GIS-based hydrologic modeling tool. In *Proceedings of the 2005 Watershed Management Conference-Managing Watersheds for Human and Natural Impacts: Engineering, Ecological, and Economic Challenges, Williamsburg, VA, USA, 19–22 July 2005*; Moglen, G.E., Ed.; American Society of Civil Engineers: Williamsburg/Alexandria, VA, USA, 2005; pp. 509–520.

59. Papathanasiou, C.; Alonistioti, D.; Kasella, A.; Makropoulos, C.; Mimikou, M. The impact of forest fires on the vulnerability of peri-urban catchments to flood events (The case of the Eastern Attica region). *Glob. NEST J. Hydrol. Water Resour.* **2012**, *14*, 294–302.

60. Papathanasiou, C.; Makropoulos, C.; Mimikou, M. Hydrological modelling for flood forecasting: Calibrating the post-fire initial conditions. *J. Hydrol.* **2015**, *529*, 1838–1850. [CrossRef]

61. Versini, P.A.; Velasco, M.; Cabello, A.; Sempere-Torres, D. Hydrological impact of forest fires and climate change in a Mediterranean basin. *Nat. Hazards* **2013**, *66*, 609. [CrossRef]

62. Nyman, P.; Sheridan, G.J.; Lane, P.N.J. Hydro-geomorphic response models for burned areas and their applications in land management. *Prog. Phys. Geogr.* **2013**, *37*, 787–812. [CrossRef]

63. Batelis, S.; Nalbantis, I. Potential effects of forest fires on streamflow in the Enipeas river basin, Thessaly, Greece. *Environ. Process.* **2014**, *1*, 73–85. [CrossRef]

64. Leopardi, M.; Scorzini, A.R. Effects of wildfires on peak discharges in watersheds. *IForest* **2015**, *8*, 302–307. [CrossRef]

65. Masek, J.G.; Lindsay, F.E.; Goward, S.N. Dynamics of urban growth in the Washington DC metropolitan area, 1973–1996, from Landsat observations. *Int. J. Remote Sens.* **2000**, *21*, 3473–3486. [CrossRef]

66. Prakash, A.; Gupta, R.P. Land-use mapping and change detection in a coal mining area-a case study in the Jharia coalfield, India. *Int. J. Remote Sens.* **1998**, *19*, 391–410. [CrossRef]

67. Psomiadis, E.; Soulis, K.; Zoka, M.; Dercas, N. Synergistic approach of Remote Sensing and GIS techniques for flash-flood monitoring and damage assessment in Thessaly plain area, Greece. *Water* **2019**, *11*, 448. [CrossRef]

68. Chatziantoniou, A.; Petropoulos, G.; Psomiadis, E. Co-orbital Sentinel 1 & 2 for wetlands mapping based on machine learning: A case study from a Mediterranean setting. *Remote Sens. Basel* **2017**, *9*, 1259. [CrossRef]

69. Mendoza, M.E.; Granados, E.L.; Geneletti, D.; Péerz-Salicrup, D.R. Analysing land cover and land use change processes at watershed level: A multitemporal study in the Lake Cuitzeo Watershed, Mexico (1975-2003). *Appl. Geogr.* **2011**, *31*, 237–250. [CrossRef]

70. Butt, A.; Shabbir, R.; Ahmad, S.S.; Aziz, N. Land use change mapping and analysis using Remote Sensing and GIS: A case study of Simly watershed, Islamabad, Pakistan. *Egypt. J. Remote Sens. Space Sci.* **2015**, *18*, 251–259. [CrossRef]

71. Behera, M.D.; Tripathi, P.; Das, P.; Srivastava, S.K.; Roy, P.S.; Joshi, C.; Behera, P.R.; Deka, J.; Kumar, P.; Khan, M.L.; et al. Remote sensing based deforestation analysis in Mahanadi and Brahmaputra river basin in India since 1985. *J. Environ. Manag.* **2018**, *206*, 1192–1203. [CrossRef]

72. Chormanski, J.; Van de Voorde, T.; De Roeck, T.; Batelaan, O.; Canters, F. Improving distributed runoff prediction in urbanized catchments with remote sensing based estimates of impervious surface cover. *Sensors* **2008**, *8*, 910–932. [CrossRef]

73. Nourani, V.; Fard, A.F.; Niazi, F.; Gupta, H.V.; Goodrich, D.C.; Kamran, K.V. Implication of remotely sensed data to incorporate land cover effect into a linear reservoir-based rainfall–runoff model. *J. Hydrol.* **2015**, *529*, 94–105. [CrossRef]

74. Vafeidis, A.T.; Drake, N.A.; Wainwright, J. A proposed method for modelling the hydrologic response of catchments to burning with the use of remote sensing and GIS. *Catena* **2007**, *70*, 396–409. [CrossRef]

75. Sun, Z.; Li, X.; Fu, W.; Li, Y.; Tang, D. Long-term effects of land use/land cover change on surface runoff in urban areas of Beijing, China. *J. Appl. Remote Sens.* **2014**, *8*, 084596. [CrossRef]

76. Tollan, A. Land-use change and floods: What do we need most, research or management? *Water Sci. Technol.* **2002**, *45*, 183–190. [CrossRef] [PubMed]

77. Savary, S.; Rousseau, A.N.; Quilbé, R. Assessing the effects of historical Land Cover Changes on runoff and low flows using remote sensing and hydrological modeling. *J. Hydrol. Eng.* **2009**, *14*, 575–587. [CrossRef]

78. Miller, J.D.; Hugh, D.S.; Welch, K.R. Using one year post-fire fire severity assessments to estimate longer-term effects of fire in conifer forests of northern and eastern California, USA. *For. Ecol. Manag.* **2016**, *382*, 168–183. [CrossRef]

79. Shi, P.-J.; Zheng, J.; Wang, J.-A.; Ge, Y.; Qiu, G.-Y. The effect of land use/cover change on surface runoff in Shenzhen region, China. *Catena* **2007**, *69*, 31–35. [CrossRef]

80. Irons, J.; Dwyer, J.L.; Barsi, J.A. The next Landsat satellite: The Landsat Data Continuity Mission. *Remote Sens. Environ.* **2012**, *122*, 11–21. [CrossRef]

81. Feranec, J.; Soukup, T.; Hazeu, G.; Jaffrain, G. (Eds.) *European Landscape Dynamics: CORINE Land Cover Data*, 1st ed.; CRC Press: Boca Raton, FL, USA, 2016.

82. Mallinis, G.; Gitas, I.Z.; Tasionas, G.; Maris, F. Multitemporal monitoring of land degradation risk due to soil loss in a fire-prone Mediterranean landscape using multi-decadal Landsat imagery. *Water Resour. Manag.* **2016**, *30*, 1255–1269. [CrossRef]

83. Efthimiou, N.; Psomiadis, E. The significance of land cover delineation on soil erosion assessment. *Environ. Manag.* **2018**, *62*, 383–402. [CrossRef]

84. Li, X.; Ling, F.; Foody, G.M.; Ge, Y.; Zhang, Y.; Du, Y. Generating a series of fine spatial and temporal resolution land cover maps by fusing coarse spatial resolution remotely sensed images and fine spatial resolution land cover maps. *Remote Sens. Environ.* **2017**, *196*, 293–311. [CrossRef]

85. Du, J.; Qian, L.; Rui, H.; Zuo, T.; Zheng, D.; Xu, Y.; Xu, C.-Y. Assessing the effects of urbanization on annual runoff and flood events using an integrated hydrological modeling system for Qinhuai River basin, China. *J. Hydrol.* **2012**, *464*, 127–139. [CrossRef]

86. Verbeiren, B.; Van De Voorde, T.; Canters, F.; Binard, M.; Cornet, Y.; Batelaan, O. Assessing urbanisation effects on rainfall-runoff using a remote sensing supported modelling strategy. *Int. J. Appl. Earth Obs.* **2013**, *21*, 92–102. [CrossRef]

87. Papathanasiou, C.; Pagana, V.; Varela, V.; Eftychidis, G.; Makropoulos, C.; Mimikou, M. *Status Survey Report for Rafina Catchment, Technical Report for Action A2: Identification of the Current Status of the Study Area*; FLIRE Project (LIFE11 ENV GR 975); National Technical University of Athens: Athens, Greece, 2013; Available online: http://www.flire.gr/wp-content/uploads/2013/06/A2-Current_status_report_update.pdf.

88. Mettos, A.; Ioakim, C.; Rondoyanni, T. Palaeoclimatic and palaeogeographic evolution of Attica-Beotia (Central-Greece). *Spec. Publ. Geol. Soc. Greece* **2000**, *9*, 187–196.

89. Kottek, M.; Grieser, J.; Beck, C.; Rudolf, B.; Rubel, F. World Map of the Köppen-Geiger climate classification updated. *Meteorol. Z.* **2006**, *15*, 259–263. [CrossRef]

90. Alexakis, D. Geochemistry of stream sediments as a tool for assessing contamination by Arsenic, Chromium and other toxic elements: East Attica region, Greece. *Eur. Water* **2008**, *21*, 57–72.

91. Efthimiou, N.; Psomiadis, E.; Panagos, P. Fire severity and soil erosion susceptibility mapping using multi-temporal earth observation data: The case of Mati fatal wildfire in eastern Attica, Greece. *Catena* **2020**, *187*, 104320. [CrossRef]

92. Soulis, K.X.; Valiantzas, J.D. Identification of the SCS-CN parameter spatial distribution using rainfall-runoff data in heterogeneous watersheds. *Water Resour. Manag.* **2013**, *27*, 1737–1749. [CrossRef]

93. Soulis, K.X.; Dercas, N.; Papadaki, C.H. Effects of forest roads on the hydrological response of a small-scale mountain watershed in Greece. *Hydrol. Process.* **2015**. [CrossRef]

94. Papathanasiou, C.; Pagana, V.; Varela, V.; Eftychidis, G.; Makropoulos, C.; Mimikou, M. *Status Survey Report for Rafina Catchment, Technical Report for Action B1: Catchment Hydrological Modelling*; FLIRE Project (LIFE11 ENV GR 975); National Technical University of Athens: Athens, Greece, 2013; Available online: http://www.flire.gr/wp-content/uploads/2013/03/B1-Catchment-Hydrological-Modelling_Report.pdf.

95. Hellenic Statistical Authority, Population and Social Conditions. Available online: https://www.statistics.gr/el/statistics/pop (accessed on 2 April 2020).

96. Xanthopoulos, G. The forest fires of 1995 and 1998 on Penteli Mountain. In Proceedings of the International Workshop on "Improving Dispatching for Forest Fire Control", Chania, Greece, 6-8 December 2002; pp. 85–94.

97. Xanthopoulos, G. Fires in the urban-wildland interface area: A complex problem. *NAGREF Q. Publ. Natl. Agric. Res. Found.* **2006**, *24*, 4–9. (In Greek)

98. Giakoumakis, S. *Floods: Alarm for the Eastern Attiki Area*; Greek Committee for Water Resources Management: Athens, Greece, 2000. (In Greek)

99. Meteoclub, History of Catastrophic Storm Events in Athens Region from 1887. Available online: http://www.meteoclub.gr/themata/egkyklopaideia/920-1887 (accessed on 2 March 2020).

100. Lazaridis, L.; Nalbantis, I. Flood protection of the Rafina torrent basin. In Proceedings of the Day conference "Flood protection of Attica", Athens, Greece, 2 November 2004. (In Greek).

101. METEONET, Hydrological Observatory of Athens, National Technical University of Athens (NTUA). Available online: http://hoa.ntua.gr/info/ (accessed on 31 March 2020).

102. Historic Floods Database of the Greek Ministry of Environment, Energy and Climate Change. Available online: http://www.ypeka.gr/en-us/Water-Resources/Floods (accessed on 31 March 2020).

103. CLC. 2018. Available online: https://land.copernicus.eu/pan-european/corine-land-cover/clc2018 (accessed on 29 January 2020).

104. NRCS (Natural Resources Conservation Service). *National Engineering Handbook, Section 4: Hydrology, Natural Resources Conservation Service*; US Department of Agriculture: Washington, DC, USA, 2004.

105. OPEKEPE Integrated Administration and Control System; Payment and Control Agency for Guidance and Guarantee Community Aid, Ministry of Agricultural Development and Food, Athens, Greece. Available online: http://www.opekepe.gr/ (accessed on 30 March 2020).

106. Panagos, P.; Van Liedekerke, M.; Jones, A.; Montanarella, L. European Soil Data Centre: Response to European policy support and public data requirements. *Land Use Policy* **2012**, *29*, 329–338. [CrossRef]

107. Orgiazzi, A.; Ballabio, C.; Panagos, P.; Jones, A.; Fernández-Ugalde, O. LUCAS Soil, the largest expandable soil dataset for Europe: A review. *Eur. J. Soil Sci.* **2018**, *69*, 140–153. [CrossRef]

108. HNMS Hellenic National Meteorological Service. Available online: http://www.hnms.gr/hnms/english/index_html (accessed on 31 March 2020).

109. Soulis, K.X.; Valiantzas, J.D. SCS-CN parameter determination using rainfall-runoff data in heterogeneous watersheds-the two-CN system approach. *Hydrol. Earth Syst. Sci.* **2012**, *16*, 1001–1015. [CrossRef]

110. Keeley, J.E. Fire intensity, fire severity and burn severity: A brief review and suggested usage. *Int. J. Wildland Fire* **2009**, *18*, 116–126. [CrossRef]

111. Athanasakis, G.; Psomiadis, E.; Chatziantoniou, A. High-resolution Earth Observation data and spatial analysis for burn severity evaluation and post-fire effects assessment in the Island of Chios, Greece. In Proceedings of the SPIE (Remote Sensing- Earth Resources and Environmental Remote Sensing/GIS Applications VII), Warsaw, Poland, 5 October 2017. SPIE, 104281P. [CrossRef]

112. Mitsopoulos, I.; Mallinis, G.; Arianoutsou, M. Wildfire risk assessment in a typical Mediterranean Wildland–Urban Interface of Greece. *Environ. Manag.* **2015**, *55*, 900–915. [CrossRef]

113. Mustafa, A.; Szydlowski, M. The impact of spatiotemporal changes in land development (1984-2019) on the increase in the runoff coefficient in Erbil, Kurdistan region of Iraq. *Remote Sens.* **2020**, *12*, 1302. [CrossRef]

114. Woodward, D.E.; Hawkins, R.H.; Jiang, R.; Hjelmfelt, A.T., Jr.; Van Mullem, J.A.; Quan, D.Q. Runoff curve number method: Examination of the initial abstraction ratio. In Proceedings of the World Water & Environmental Resources Congress 2003 and Related Symposia, Philadelphia, PA, USA, 23–26 June 2003. [CrossRef]

115. Mishra, S.K.; Singh, V.P. A relook at NEH-4 curve number data and antecedent moisture condition criteria. *Hydrol. Process.* **2006**, *20*, 2755–2768. [CrossRef]

116. Elhakeem, M.; Papanicolaou, A.N. Estimation of the runoff curve number via direct rainfall simulator measurements in the state of Iowa, USA. *Water Resour. Manag.* **2009**, *23*, 2455–2473. [CrossRef]

117. McCuen, R.H. Approach to confidence interval estimation for curve numbers. *J. Hydrol. Eng.* **2002**, *7*, 43–48. [CrossRef]

118. Ponce, V.M.; Hawkins, R.H. Runoff curve number: Has it reached maturity? *J. Hydrol. Eng.* **1996**, *1*, 11–18. [CrossRef]

119. Hawkins, R.H. Runoff curve numbers for partial area watersheds. *J. Irrig. Drain. Div. ASCE* **1979**, *105*, 375–389.

120. Grove, M.; Harbor, J.; Engel, B. Composite vs. Distributed curve numbers: Effects on estimates of storm runoff depths. *J. Am. Water Resour. Assoc.* **1998**, *34*, 1015–1023. [CrossRef]

121. Lantz, D.G.; Hawkins, R.H. Discussion of "Long-Term Hydrologic Impact of Urbanization: A Tale of Two Models". *J. Water Res. Plan. ASCE* **2001**, *127*, 13–19.

122. Higginson, B.; Jarnecke, J. *Salt Creek BAER-2007 Burned Area Emergency Response. Hydrology Specialist Report (Unnumbered)*; U.S. Department of Agriculture, Forest Service, Intermountain Region, Unita National Forest: Provo, UT, USA, 2007; pp. 1–11.

123. Kinoshita, A.M.; Hogue, T.S.; Napper, C.A. *Guide for Pre- and Postfire Modeling and Application in the Western United States*; United States Department of Agriculture: Washington, DC, USA, 2013. Available online: http://bluegold.sdsu.edu/docs/2014_KinoshitaHogueNapper_GuideforPreandPostfireModeling.pdf (accessed on 2 March 2020).

124. Cerrelli, G.A. FIRE HYDRO: A simplified method for predicting peak discharges to assist in the design of flood protection measures for western wildfires. In *Proceedings of the 2005 Watershed Management Conference-Managing Watersheds for Human and Natural Impacts: Engineering, Ecological, and Economic Challenges, Williamsburg, VA, USA, 19–22 July 2005*; Moglen, G.E., Ed.; American Society of Civil Engineers: Williamsburg/Alexandria, VA, USA, 2005; pp. 935–941.

125. Springer, E.P.; Hawkins, R.H. Curve number and peakflow responses following the Cerro Grande fire on a small watershed. In *Proceedings of the 2005 Watershed Management Conference-Managing Watersheds for Human and Natural Impacts: Engineering, Ecological, and Economic Challenges. Williamsburg, VA, USA, 19–22 July 2005*; Moglen, G.E., Ed.; American Society of Civil Engineers: Williamsburg/Alexandria, VA, USA, 2005; pp. 459–470.

126. Foltz, R.B.; Robichaud, P.R.; Rhee, H.A. *Synthesis of Post-Fire Road Treatments for BAER Teams: Methods, Treatment Effectiveness, and Decision Making Tools for Rehabilitation*; General Technical Report RMRS-GTR-228; U.S. Department of Agriculture, Forest Service, Rocky Mountain Research Station: Fort Collins, CO, USA, 2009; p. 152.

127. NRCS. *Hydrologic Analyses of Post-Wildfire Conditions, DRAFT Technical Note*; Natural Resources Conservation Service, USDA: Washington, DC, USA, 2015. Available online: https://directives.sc.egov.usda.gov/OpenNonWebContent.aspx?content=37227.wba (accessed on 31 March 2020).

128. Canfield, H.E.; Goodrich, D.C.; Burns, I.S. Selection of parameters values to model post-fire runoff and sediment transport at the watershed scale in southwestern forests. In *Proceedings of the 2005 Watershed Management Conference-Managing Watersheds for Human and Natural Impacts: Engineering, Ecological, and Economic Challenges, Williamsburg, VA, USA, 19–22 July 2005*; Moglen, G.E., Ed.; American Society of Civil Engineers: Williamsburg/Alexandria, VA, USA, 2005; pp. 561–572.

129. Flood Risk Management Plan of the Attica Water District, Energy and Climate Change. Available online: https://floods.ypeka.gr/index.php?option=com_content&view=article&id=272&Itemid=630 (accessed on 3 April 2020).

130. Buhk, C.; Götzenberger, L.; Wesche, K.; Sánchez Gómez, P.; Hensen, I. Post-fire regeneration in a Mediterranean pine forest with historically low fire frequency. *Acta Oecol.* **2006**, *30*, 288–298. [CrossRef]

131. Mouillot, F.; Rambal, S.; Joffre, R. Simulating climate change impacts in fire frequency and vegetation dynamics in a Mediterranean-type ecosystem. *Glob. Chang. Biol.* **2002**, *8*, 423–437. [CrossRef]

132. Price, O.; Bradstock, R. Countervailing effects of urbanization and vegetation extent on fire frequency on the wildland urban interface: Disentangling fuel and ignition effects. *Landsc. Urban Plan.* **2014**, *130*, 81–88. [CrossRef]

133. Soulis, K.X.; Ntoulas, N.; Nektarios, P.A.; Kargas, G. Runoff reduction from extensive green roofs having different substrate depth and plant cover. *Ecol. Eng.* **2017**, *102*, 80–89. [CrossRef]

MDPI

St. Alban-Anlage 66

4052 Basel

Switzerland

Tel. +41 61 683 77 34

Fax +41 61 302 89 18

www.mdpi.com

Water Editorial Office

E-mail: water@mdpi.com

www.mdpi.com/journal/water